Analyzing Baseball Data with R

"Our community has continued to grow exponentially, thanks to those who inspire the next generation. And inspiring the next generation is what the authors of Analyzing Baseball Data with R are doing. They are setting the career path for still thousands more. We all need some sort of kickstart to take that first or second step. You may be a beginner R coder, but you need access to baseball data. How do you access this data, how do you manipulate it, how do you analyze it? This is what this book does for you. But it does more, by doing what sabermetrics does best: it asks baseball questions. Throughout the book, baseball questions are asked, some straightforward, and others more thought-provoking."

-From the Foreword by Tom Tango

Analyzing Baseball Data with R Third Edition introduces R to sabermetricians, baseball enthusiasts, and students interested in exploring the richness of baseball data. It equips you with the necessary skills and software tools to perform all the analysis steps, from importing the data to transforming them into an appropriate format to visualizing the data via graphs to performing a statistical analysis.

The authors first present an overview of publicly available baseball datasets and a gentle introduction to the type of data structures and exploratory and data management capabilities of R. They also cover the ggplot2 graphics functions and employ a tidyverse-friendly workflow throughout. Much of the book illustrates the use of R through popular sabermetrics topics, including the Pythagorean formula, runs expectancy, catcher framing, career trajectories, simulation of games and seasons, patterns of streaky behavior of players, and launch angles and exit velocities. All the datasets and R code used in the text are available for download online.

New to the third edition is the revised R code to make use of new functions made available through the tidyverse. The third edition introduces three chapters of new material, focusing on communicating results via presentations using the Quarto publishing system, web applications using the Shiny package, and working with large data files. An online version of this book is hosted at https://beanumber.github.io/abdwr3e/.

Jim Albert is a Distinguished University Professor of Statistics at Bowling Green State University. He has authored or co-authored several books including Curve Ball and Visualizing Baseball and was the editor of the Journal of Quantitative Analysis of Sports. He received the Significant Contributor to Statistics in Sports award in 2003 from the Section of Statistics in Sports of the American Statistical Association.

Ben Baumer is a Professor of Statistical and Data Sciences at Smith College. Previously a statistical analyst for the New York Mets, he is a co-author of The Sabermetric Revolution and Modern Data Science with R. He has received the Waller Education Award from the ASA Section on Statistics and Data Science Education, the Significant Contributor Award from the ASA Section on Statistics in Sports, and the Contemporary Baseball Analysis Award from the Society for American Baseball Research.

Max Marchi is a Baseball Analytics Analyst for the Cleveland Indians. He was a regular contributor to The Hardball Times and Baseball Prospectus websites and previously consulted for other MLB clubs.

Chapman & Hall/CRC
The R Series

Series Editors
John M. Chambers, Department of Statistics, Stanford University, California, USA
Torsten Hothorn, Division of Biostatistics, University of Zurich, Switzerland
Duncan Temple Lang, Department of Statistics, University of California, Davis, USA
Hadley Wickham, RStudio, Boston, Massachusetts, USA

Recently Published Titles

A Criminologist's Guide to R: Crime by the Numbers
Jacob Kaplan

Analyzing US Census Data: Methods, Maps, and Models in R
Kyle Walker

ANOVA and Mixed Models: A Short Introduction Using R
Lukas Meier

Tidy Finance with R
Christoph Scheuch, Stefan Voigt, and Patrick Weiss

Deep Learning and Scientific Computing with R torch
Sigrid Keydana

Model-Based Clustering, Classification, and Density Estimation Using mclust in R
Lucca Scrucca, Chris Fraley, T. Brendan Murphy, and Adrian E. Raftery

Spatial Data Science: With Applications in R
Edzer Pebesma and Roger Bivand

Modern Data Visualization with R
Robert Kabacoff

Learn R: As a Language, Second Edition
Pedro J. Aphalo

Spatial Analysis in Geology Using R
Pedro M. Nogueira

Analyzing Baseball Data with R, Third Edition
Jim Albert, Benjamin S. Baumer and Max Marchi

For more information about this series, please visit: https://www.crcpress.com/
Chapman--HallCRC-The-R-Series/book-series/CRCTHERSER

Analyzing Baseball
Data with R
Third Edition

Jim Albert, Benjamin S. Baumer and Max Marchi

CRC Press
Taylor & Francis Group
Boca Raton London New York

CRC Press is an imprint of the
Taylor & Francis Group, an **informa** business

A CHAPMAN & HALL BOOK

Designed cover image: © Jim Albert, Benjamin S. Baumer and Max Marchi

First edition published 2013
Second edition published 2019
Third edition published 2025
by CRC Press
2385 NW Executive Center Drive, Suite 320, Boca Raton FL 33431

and by CRC Press
4 Park Square, Milton Park, Abingdon, Oxon, OX14 4RN

CRC Press is an imprint of Taylor & Francis Group, LLC

ISBN: 978-1-0326-68154 (hbk)
ISBN: 978-1-0326-68093 (pbk)
ISBN: 978-1-0326-68239 (ebk)

DOI: 10.1201/9781032668239

Typeset in CMR10
by KnowledgeWorks Global Ltd

Publisher's note: This book has been prepared from camera-ready copy provided by the authors.

Contents

Foreword

Back in the late 1980s, as a college student, I would go to the NHL offices in downtown Montreal and pick up their official end-of-season statistics package. This was something that was reserved for the media, but somehow I had the idea to ask, and they were kind enough to oblige. Since I was learning dBase, I then had to manually enter every piece of data in that package, probably some 20,000 discrete data values. Combining sports, numbers, and computers was a labor of love for me. In the 1980s, Pete Palmer and Bill James inspired the sabermetric revolution: my career path was set, as was that of thousands more.

Our community has continued to grow exponentially, thanks to those who inspire the next generation. And inspiring the next generation is what the authors of *Analyzing Baseball Data with R* are doing. They are setting the career path for still thousands more. We all need some sort of kickstart to take that first or second step. You may be a beginner R coder, but you need access to baseball data. How do you access this data, how do you manipulate it, how do you analyze it? This is what this book does for you. But it does more, by doing what sabermetrics does best: it asks baseball questions. Throughout the book, baseball questions are asked, some straightforward and others more thought-provoking. Either way, the tools are introduced, whether plotting data or performing calculations in order to answer those questions. And as good sabermetrics does, each answer will make you ask two more questions.

In addition to being an ideal reference in book form, the authors generously make the content and data readily accessible online, through GitHub as well as in blog form. Not only are the references ideal, but they are also welcoming to new readers. You will find nuggets throughout to get you started. And more than anything, with Jim, Max, and Ben, you will simply find good people. Having the next generation being inspired by talented and good folks is all that we can ask for. And they deliver.

Tom Tango

Senior Data Architect, Major League Baseball

Preface

What's New in the Third Edition?

In the ten years since the publication of the first edition of this book, there have been many new developments in R, including the introduction of many new packages. (As we are writing this preface, there are currently 20,122 packages available on the CRAN package repository.) In particular, the **tidyverse** collection of packages has streamlined workflows for data visualization and data manipulation. In this third edition, we have revised the R code to embrace the new functions and paradigms available through the **tidyverse**. This includes porting the book's source code from LaTeX to Quarto, enabling us to simultaneously maintain an online web version of the book at: https://beanumber.github.io/abdwr3e/.

This third edition introduces three chapters with new material. One important aspect of working as a baseball analyst is communicating one's findings to other people in the organization. Chapters 14 and 15 focus on communicating results via presentations using the Quarto publishing system and web applications using the Shiny package. Given the availability of large quantities of baseball data, one challenge is how to efficiently work with these data. Chapter 12 explores methods for downloading, storing, retrieving, and analyzing large Statcast datasets. The "Batted Ball Data from Statcast" chapter from the previous revision of the book has been rewritten in Chapter 13 to focus on the interesting pattern of home run hitting during the Statcast era. Appendices A, B, and C have been revised to reflect new realities, most notably including new functionality in the **baseballr** package and the disappearance of the PITCHf/x data source.

Preface from the First Edition

Baseball has always had a fascination with statistics. Schwarz (2004) documents the quantitative measurements of teams and players since the beginning of professional baseball history in the 19th century. Since the foundation of the Society for American Baseball Research (SABR) in 1971, an explosion of new measures have been developed for understanding offensive and defensive contributions of players. One can learn much about the current developments in sabermetrics by viewing articles at websites such as https://www.baseballprospectus.com, https://www.hardballtimes.com, and https://www.fangraphs.com.

The quantity and granularity of baseball data have exhibited remarkable growth since the birth of the Internet. The first data were collected for players and teams for individual seasons—these data were what would be displayed on the back side of a Topps baseball card. The volunteer-run Project Scoresheet organized the collection of play-by-play game data, and these data are currently freely available at the Retrosheet organization at https://www.retrosheet.org. Since 2006, PITCHf/x data has measured the speed and trajectory of every pitched ball, and since 2015, Statcast has collected the speeds and locations of batted balls and the locations and movements of baserunners and fielders at fractions of a second.

The ready availability of these large baseball datasets has led to challenges for baseball enthusiasts interested in answering baseball questions with these data. It can be problematic to download and organize the data. Standard statistical software packages may be well-suited for working with small datasets of a specific format, but they are less helpful in merging datasets of different types or performing particular types of analyses, say contour graphs of pitch locations, that are helpful for PITCHf/x data.

Fortunately, a new open-source statistical computing environment, R, has experienced increasing popularity within the statistics, data science, and computer science communities. R is a system for statistical computation and graphics, and it is a computer language designed for typical and possibly specialized statistical and graphical applications. The software is available for Linux, Windows, and Macintosh platforms from http://www.r-project.org.

The public availability of baseball data and the open-source R software is an attractive marriage. R provides a large range of tools for importing, arranging, and organizing large datasets. Through the use of built-in functions and collections of packages from the R user-community, one can perform various data and graphical analyses, and communicate this work easily to other baseball enthusiasts over the Internet. In 2014, one of us asked a number of MLB team analytics groups about their use of R and here are some responses:

- "We use: R, MySQL / Oracle, Perl, PHP".
- "We do use R extensively, and it is our primary statistical package. The only other major tool we use is probably Excel".
- "We do use R here. It is our primary statistical package for projects that need something more than the statistical functions in Excel".

- "With the occasional exception of Python+NumPy, R is the only statistical programming language or package we use".
- "We do use R. It's used in conjunction with Excel for analysis".

It is clear that R is a major tool for the analytical work of MLB teams.

The purpose of this book is to introduce R to sabermetricians, baseball enthusiasts, and students interested in exploring baseball data.

Overview of Chapters

The contents of this book can be divided into three themes: chapters devoted to popular topics within sabermetrics, chapters focusing on particular datasets, and chapters that illustrate R tools.

Sabermetrics

- Chapter 4: The Relation Between Runs and Wins
- Chapter 5: Value of Plays Using Run Expectancy
- Chapter 6: Balls and Strikes Effects
- Chapter 7: Catcher Framing
- Chapter 8: Career Trajectories
- Chapter 9: Simulation
- Chapter 10: Exploring Streaky Performances

Baseball Data Sets

- Chapter 1: The Baseball Datasets
- Chapter 13: Home Run Hitting
- Appendix A: Retrosheet Files Reference
- Appendix B: Historical notes on PITCHf/x
- Appendix C: Statcast Data Reference

R tools

- Chapter 2: Introduction to R
- Chapter 3: Graphics
- Chapter 11: Using a Database to Compute Park Factors
- Chapter 12: Working with Large Data
- Chapter 14: Making a Scientific Presentation Using Quarto
- Chapter 15: Using Shiny for Baseball Applications

Two fundamental ideas in sabermetrics are the relationship between runs and wins, and the measurement of the value of baseball events by runs. Chapter 4 explores the famous Pythagorean formula derived by Bill James, and Chapters 5 and 6 describe the value of plays and pitch sequences using run expectancy.

It is fascinating to explore career performance trajectories of ballplayers, and Chapter 8 illustrates the use of R to fit quadratic models to player trajectories. Chapter 9 illustrates the use of R simulation functions to simulate a game of baseball by a Markov chain model and simulate a season of baseball competition. Baseball fans are interested in streaky patterns of performance of teams and players and Chapter 10 explores methods of describing and understanding the significance of streaky patterns of hitting.

Chapter 1 provides an overview of the publicly available baseball datasets and Chapter 13 describes many of the new variables available in the Statcast system. The datafiles available through Retrosheet (Appendix A), MLBAM Gameday (Appendix B), and Statcast (Appendix C) are relatively sophisticated, so we provide detailed descriptions for downloading and reading these data into R.

Chapter 2 gives a gentle introduction to the type of data structures and exploratory and data management capabilities of R. One of the strongest features of R is its graphics capabilities—Chapter 3 provides an overview of the **ggplot2** graphics package. Given the large size of baseball datasets, it may be more convenient to work with a relational database and Chapter 11 illustrates the application of several R packages to interface with a MySQL database. This material motivates a discussion about issues working with large datasets and additional technologies in Chapter 12. The book concludes in Chapters 14 and 15 by describing tools for communicating results of baseball work.

How to Use this Book

We encourage the reader to work on the book datasets and try out the presented R code as the chapters are read. All of the small data files and R code used in the book are available at the GitHub repository for the associated R package **abdwr3edata** (http://github.com/beanumber/abdwr3edata). In addition, at the "Exploring Baseball Data with R" book blog at https://baseballwithr.wordpress.com, these authors and others provide advice on using R in sabermetrics research and keep the reader informed of new developments in R software and baseball datasets.

There is an active academic research community in baseball as demonstrated by published referred articles in journals, particularly *The Journal of Quantitative Analysis in Sports (JQAS)* and the *Journal of Sports Analytics*. The recently published articles Brill, Deshpande, and Wyner (2023), Gerber and Craig (2021), Bouzarth et al. (2021), Hirotsu and Bickel (2019), and Healey (2019) in *JQAS* describe work on pitcher fatigue, prediction of future performance, proper defensive positioning, measuring the value of the sacrifice bunt, and measuring the value of a pitch. Reading these articles and attending sports analytics conferences (e.g., the New England Symposium on Statistics in Sports or the Carnegie Mellon Sports Analytics Conference) are great ways to deepen your knowledge. Recent work in sports analytics more broadly includes a new CRAN Task View for Sports Analytics that includes many of the R packages used in this book, a systematic review of these packages and their properties (Casals et al. 2023), and an attempt to connect big ideas (many of which originated in baseball and are described in this book) across various sports (Benjamin S. Baumer, Matthews, and Nguyen 2023).

We imagine this book as a first step toward a professional career in baseball analytics. Other stops along the path to professionalization might include the SABR Analytics Certification courses. Three levels are offered, with the highest level presenting R programming material consistent with what appears in this book.

Acknowledgments

We have appreciated all of the positive comments and suggestions on the first two editions. We're especially grateful to Jason Osborne and a slew of GitHub users for catching errors in the previous editions that we were able to correct in this edition. We believe the book is useful for quantitatively oriented baseball fans who would like to learn R to perform their own analyses. We agree with Donoho (2017) that a careful study of Tango, Lichtman, and Dolphin (2007) and this book would be an excellent introduction to data preparation and exploration within the context of baseball.

The authors are very grateful for the efforts of our editors, John Kimmel and David Grubbs, who played an important role in our collaboration and provided us with timely reviews that led to significant improvements of the manuscript. We wish to thank our partners, Anne, Ramona, and Cory, and our children, Lynne, Bethany, Steven, Alice, and Arlo for their encouragement and inspiration. Although the three of us live thousands of miles apart, we share a passion for statistics, baseball, and the knowledge that one can learn about the game through the exploration of data.

Northampton, MA (and Medellín, Colombia) and Findlay, OH

December 2023

About this Book

About the Authors

Jim Albert is Emeritus Distinguished University Professor, Department of Mathematics and Statistics at Bowling Green State University. Jim is author of *Teaching Statistics Using Baseball*, *Visualizing Baseball*, and coauthor (with Jay Bennett) of *Curve Ball*. Jim received the Significant Contributor to Statistics in Sports in 2003, an award given by the Section of Statistics in Sports of the American Statistical Association.

Max Marchi is a Baseball Analytics Analyst for the Cleveland Guardians. Max was a regular contributor to *The Hardball Times* and *Baseball Prospectus* websites and previously consulted for other MLB clubs.

Benjamin S. Baumer is a professor in the Statistical & Data Sciences program at Smith College. He has been a practicing data scientist since 2004, when he became the first full-time statistical analyst for the New York Mets. Ben is a co-author of *Modern Data Science with R*, *The Sabermetric Revolution*, and *Analyzing Baseball Data with R*. Ben has received the Waller Education Award from the ASA Section on Statistics and Data Science Education, the Significant Contributor Award from the ASA Section on Statistics in Sports, and the Contemporary Baseball Analysis Award from the Society for American Baseball Research.

About our Computing Environment

We used R version 4.3.2 (R Core Team 2023) and the following R packages: abdwr3edata v. 0.0.2 (Benjamin S. Baumer and Albert 2024), arrow v. 14.0.0.2 (Richardson et al. 2023), baseballr v. 1.6.0 (Petti and Gilani 2024), bench v. 1.1.3 (Hester and Vaughan 2023), broom v. 1.0.5 (Robinson, Hayes, and Couch 2023), dbplyr v. 2.4.0 (Wickham, Girlich, and Ruiz 2023), downlit v. 0.4.3 (Wickham 2023a), dplyr v. 1.1.4 (Wickham et al. 2023), duckdb v. 0.9.2.1 (Mühleisen and Raasveldt 2023), fs v. 1.6.3 (Hester, Wickham, and Csárdi 2023), ggplot2 v. 3.4.4 (Wickham 2016a), ggrepel v. 0.9.5 (Slowikowski 2024), grateful v. 0.2.4 (Francisco Rodriguez-Sanchez and Connor P. Jackson 2023), here v. 1.0.1 (Müller 2020), Hmisc v. 5.1.1 (Harrell Jr 2023), kableExtra v. 1.3.4 (Zhu 2024), knitr v. 1.45 (Xie 2014, 2015, 2023), Lahman v. 11.0.0 (Friendly et al. 2023), latex2exp v. 0.9.6 (Meschiari 2022), LearnBayes v. 2.15.1 (Albert 2018), lme4 v. 1.1.35.1 (Bates et al. 2015), lobstr v. 1.1.2 (Wickham 2022a), lubridate v. 1.9.3 (Grolemund and Wickham 2011), mdsr v. 0.2.7 (Benjamin S. Baumer, Kaplan, and Horton 2021a), metR v. 0.14.1 (Campitelli 2021), mgcv v. 1.9.1 (S. N. Wood 2003, 2004, 2011; S. N. Wood et al. 2016; S. N. Wood 2017), mlbplotR v. 1.1.0 (Carl and Kay 2023), modelr v. 0.1.11 (Wickham 2023b), patchwork v. 1.2.0 (Pedersen 2024), readr v. 2.1.5 (Wickham, Hester, and Bryan 2024), remotes v. 2.4.2.1 (Csárdi et al. 2023), RMariaDB v. 1.3.1 (Müller et al. 2023), rmarkdown v. 2.25 (Xie, Allaire, and Grolemund 2018; Xie, Dervieux, and Riederer 2020; Allaire et al. 2023), RSQLite v. 2.3.4 (Müller et al. 2024), rvest v. 1.0.3 (Wickham 2022b), shiny v. 1.8.0 (Chang et al. 2023), skimr v. 2.1.5 (Waring et al. 2022), stringr v. 1.5.1 (Wickham 2023c), tidyverse v. 2.0.0 (Wickham et al. 2019), xml2 v. 1.3.6 (Wickham, Hester, and Ooms 2023), xtable v. 1.8.4 (Dahl et al. 2019), zoo v. 1.8.12 (Zeileis and Grothendieck 2005).

Many of the data graphics in this book use a specific shade of blue used by CRC Press and denoted by the variable `crcblue` in our R code. To make your graphs match ours, you will need to define `crcblue`.

```
crcblue <- "#2905a1"
```

In this full–color version of the book, we also use a pre-defined color-blind-safe diverging color palette.

```
crc_fc
```

```
[1] "#2905a1" "#e41a1c" "#4daf4a" "#984ea3"
```

Our working directory is set using the `here()` function from the **here** package. In that directory are three subdirectories that are referenced in our code: `data`, `data_large`, and `scripts`. The `data` directory contains small data files that are available on our GitHub repository, as are the R scripts in `scripts`. However, while the data in `data_large` is necessary to compile the book, it is too big to host in the GitHub repository. Instructions for creating these data files locally appear in relevant places in the book, most notably Chapter 12 and Appendix A.

1

The Baseball Datasets

1.1 Introduction

Baseball's marriage with numbers goes back to the origins of the sport. When the first box scores and the first stats appeared in newspapers in the 1840s, the pioneers of the game had not yet decided the ultimate distance between the pitcher's rubber and home plate, nor the number of balls needed to be awarded a base.

This chapter introduces four rich sources of freely available baseball data: the Lahman database, Retrosheet, PITCHf/x, and Statcast via Baseball Savant. Baseball records from these sources have a growing level of detail, from seasonal stats available since the 1871 season, to box score data for individual games, to play-by-play accounts covering most games since 1913, to extremely detailed pitch-by-pitch data recorded for nearly all the pitches thrown in Major League Baseball parks since 2008, to player tracking data recorded every fifteenth of a second since 2015. Examples throughout this book will predominately use subsets of data coming from these four sources.

1.2 The Lahman Database: Season-by-Season Data

1.2.1 Bonds, Aaron, Ruth, and Rodriguez home run trajectories

In the 2007 baseball season, Barry Bonds became the new home run king, surpassing Hank Aaron's record of 755 career home runs. Aaron had held the throne since 1974 when he had moved past the legendary Babe Ruth with his 715th home run. In recent years, Alex Rodriguez was believed to have a great chance of breaking Bonds' record. Figure 1.1 plots the cumulative home runs of Bonds, Aaron, Ruth, and Rodriguez as a function of their age. It is clear from the graph that the home run trajectories of the four sluggers have followed different paths. Rodriguez was the clear home run leader—followed by Aaron—through age 35. Aaron and Ruth had similar career home run paths until retirement. Bonds was far behind Aaron and Ruth in career home runs

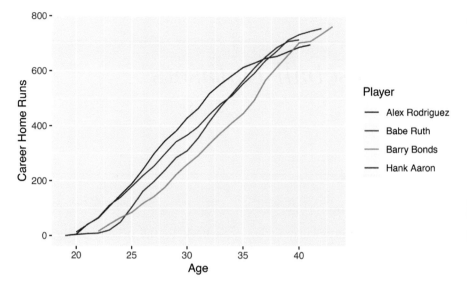

FIGURE 1.1
Career home runs by age for the top four home run hitters in baseball history.

in his 30s, but narrowed the gap and overtook the other sluggers in his 40s. Rodriguez' home run production slowed down in the final years of his career.

Babe Ruth began his career as a teenage pitcher for the Boston Red Sox in the so-called Deadball Era when home runs were rare. Ruth's home run impact was not felt until his sixth season, when he began sending the ball out of the park with regularity and outslugged nearly every other American League team with 29 home runs. Given his late start, his career line is S-shaped due to his slow start and inevitable decline at the end of his career.

Hank Aaron also made his MLB debut at a very young age and shows a nearly straight line in the graph for the best part of his career. His pattern of hitting home runs was marked by consistency as he hit between 30 and 50 home runs for most seasons of his career. Similar to the Babe, Aaron also declined in the final years of his career, hitting 20, 12, and 10 home runs from 1974 to 1976.

Barry Bonds had a relatively late major league debut as he did not come to an agreement with the team that first drafted him and was not in the career home run race until after his 35th birthday. Toward the end of his career, Bonds put together impressive season home run counts of 49, 73, 46, 45, and 45 home runs, closing in on Ruth's 714 mark. Then, after missing most of the 2005 season because of injuries, he completed the chase to the record with two solid seasons (26 and 28 homers) when he was 42 and 43 years old.

Alex Rodriguez debuted as a shortstop for the Seattle Mariners when he was 18 years old. He was a prolific home run hitter in the early part of his career, hitting over 400 home runs before the age of 30. His home run production slowed down in his mid-30s due to injuries and his suspension during the 2014 season for his role in the Biogenesis scandal.

To compare sluggers, a researcher needs season-to-season batting data including age and home run counts for Bonds, Aaron, Ruth, and Rodriguez. One needs these data for a wide range of seasons, as Ruth's career began in 1914 and Rodriguez' career ended in 2016.

For many years database journalist and author Sean Lahman has been making available at his website[1] a database containing pitching, hitting, and fielding statistics for the entire history of professional baseball from 1871 to the current season (Lahman 2018). The data are available in several formats, including a set of comma-separated-value (CSV) files that we used in the first edition of this book. The **Lahman** package now provides these data to R directly, obviating the need to download the CSVs. There is a one-to-one relationship between the CSV files and the data frames available through the **Lahman** package. We will focus our discussion on the tables available through the **Lahman** package.

1.2.2 Obtaining the database

To install the **Lahman** package, simply execute the following command.

```
install.packages("Lahman")
```

In addition, several vignettes that explain more about how to use the package are included, and the original sources can be found at https://github.com/cdalzell/Lahman. One is encouraged to read the documentation provided in the vignettes to learn about the contents of these files. For example, the following code will pull up the introductory vignette in RStudio.

```
vignette("vignette-intro", package = "Lahman")
```

Here we give a general description of the variables in the data tables most relevant for the studies described in this book.

1.2.3 The People table

The `People` table is a registry of baseball people. It contains bibliographic information on every player and manager who have appeared at the Major League Baseball level and of all people who have been inducted into the

---

[1]http://seanlahman.com/

TABLE 1.1

Tables in the Lahman database.

File	Description
AllStarFull	Players' appearances in All-Star Games
Appearances	Seasonal players' appearances by position
AwardsManagers	Recipients of the Manager of the Year Award
AwardsPlayers	Players recipients of the various Awards
AwardsShareManagers	Voting results for the Manager of the Year Award
AwardsSharePlayers	Voting results for the various Awards for players
Batting	Seasonal batting statistics
BattingPost	Seasonal batting statistics for post-season
Fielding	Seasonal fielding statistics
FieldingOF	Seasonal appearances at the three outfield positions
FieldingPost	Seasonal fielding data for post-season
HallOfFame	Voting results for the Hall of Fame
Managers	Seasonal data for managers
ManagersHalf	Seasonal split data for managers
People	Biographical information for individuals appearing in the database
Pitching	Seasonal pitching statistics
PitchingPost	Seasonal pitching statistics for post-season
Salaries	Seasonal salaries for players
Schools	List of college teams
SchoolsPlayers	Information on schools attended by players
SeriesPost	Outcomes of post-season series
Teams	Seasonal stats for teams
TeamsFranchises	Timelines of Franchises
TeamsHalf	Seasonal split stats for teams

Baseball Hall of Fame.[2] Each row of the `People` table constitutes a short biography of a person, reporting on dates and places of birth and death, height and weight, throwing hand and batting side, and the dates of the first and last game played.

---

[2]Examples of people who never played Major League Baseball but have been inducted into the Hall of Fame (therefore having an entry in the `People` table) are baseball pioneer Henry Chadwick and career Negro Leaguer Josh Gibson.

TABLE 1.2
First row of the `People.csv` file.

field_name	Value
playerID	aardsda01
birthYear	1981
birthMonth	12
birthDay	27
birthCountry	USA
birthState	CO
birthCity	Denver
deathYear	NA
deathMonth	NA
deathDay	NA
deathCountry	NA
deathState	NA
deathCity	NA
nameFirst	David
nameLast	Aardsma
nameGiven	David Allan
weight	215
height	75
bats	R
throws	R
debut	2004-04-06
finalGame	2015-08-23
retroID	aardd001
bbrefID	aardsda01
deathDate	NA
birthDate	1981-12-27

Players are identified throughout the pitching, batting, and fielding tables in the Lahman's database by an id code, and the `People` table is useful for retrieving the name of the player associated with a particular identifier. The table also reports player identification codes of other databases, in particular the ones used by Retrosheet, so one can link players from the Lahman and Retrosheet databases.

For illustration purposes, we display below the header and first row of the `People` table which gives information about the first player in the database: David Aardsma. For clarity, we place Aardsma's information in a table format in Table 1.2.

From this information, we learn some details about Aardsma's life. David Aardsma was born on December 27, 1981 in Denver, Colorado. Aardsma weighed 215 pounds and was 75 inches tall. He threw and batted right-handed, and he played in the big leagues from April 6, 2004 to August 23, 2015. There are a series of blank columns corresponding to death information, which is obviously unavailable for a living person. Finally, there are various identifying codes for the player. The value of `playerID`, `aardsda01`, is the identifying code for David Aardsma in every table in the Lahman's database. The value of the variable `retroID`, `aardd001`, is the player id specific to the Retrosheet files to be described in Section 1.3.

1.2.4   The Batting table

The `Batting` table contains all players' batting statistics by season and team from 1871 to the present season. Players in this table are identified with their `playerID`; for example, the season batting statistics of Hank Aaron appear in this table with the identification `playerID = aaronha01`. Each row of the table contains the statistics compiled by a player, during a single season (variable `yearID`), for a particular team (variable `teamID`).

Players who changed teams during a particular season have multiple rows for the season. The `stint` variable indicates the order in which the player moved between teams. For example, Lou Brock, who moved during the 1964 season from the Chicago Cubs to the St. Louis Cardinals, has the following batting rows for the 1964 season.

```
# A tibble: 2 x 22
  playerID  yearID stint teamID lgID       G    AB     R     H
  <chr>      <int> <int> <fct>  <fct>  <int> <int> <int> <int>
1 brocklo01   1964     1 CHN    NL        52   215    30    54
2 brocklo01   1964     2 SLN    NL       103   419    81   146
# i 13 more variables: X2B <int>, X3B <int>, HR <int>,
#   RBI <int>, SB <int>, CS <int>, BB <int>, SO <int>,
#   IBB <int>, HBP <int>, SH <int>, SF <int>, GIDP <int>
```

Batting statistics variables are identified by their traditional abbreviations such as AB, R, H, 2B, etc., so the column names of the batting tables should be easily understood by those familiar with baseball box scores. Note that R does not allow object names that start with numbers, so the "2B" column in `Batting` is called X2B in the `Batting` data frame. If one has questions about the meaning of one particular column name, the documentation with the package gives the variable descriptions.

An excerpt of the `Batting` table for Babe Ruth is conveniently formatted in Table 1.3. This table shows his batting statistics for his early seasons as a Boston Red Sox pitcher, his years for the Yankees when he became a great home run slugger, and his seasons at the twilight of his career with the Boston Braves.

TABLE 1.3
Batting statistics for Babe Ruth, taken from the `Batting` table.

yearID	teamID	AB	H	HR
1914	BOS	10	2	0
1915	BOS	92	29	4
1916	BOS	136	37	3
1917	BOS	123	40	2
1918	BOS	317	95	11
1919	BOS	432	139	29
1920	NYA	457	172	54
1921	NYA	540	204	59
1922	NYA	406	128	35
1923	NYA	522	205	41
1924	NYA	529	200	46
1925	NYA	359	104	25
1926	NYA	495	184	47
1927	NYA	540	192	60
1928	NYA	536	173	54
1929	NYA	499	172	46
1930	NYA	518	186	49
1931	NYA	534	199	46
1932	NYA	457	156	41
1933	NYA	459	138	34
1934	NYA	365	105	22
1935	BSN	72	13	6

Only count statistics such as the count of at-bats and count of hits are reported in the batting table. Derived statistics such as a batting average need to be computed from these count statistics. For example, a researcher who wants to know Ruth's batting average for the 1919 season has to calculate it following paragraph 10.21(b) of the *Official Baseball Rules* (Official Playing Rules Committee 2018) that instructs to "divide the number of safe hits by the total times at bat". The relevant columns are H and AB, and the desired result is 139 / 432 = .322. Some statistics are not visible for Babe Ruth as they were not recorded in the 1920s. For example, the counts of intentional walks (IBB) are blank for Ruth's seasons, indicating that they were not recorded.

1.2.5 The Pitching table

The `Pitching` table contains season-by-season pitching data for players. This table contains the traditional count data for pitching such as W (number of wins), L (number of losses), G (games played), BB (number of walks), and SO (number of strikeouts). In addition, this dataset contains several derived

TABLE 1.4

Pitching statistics for Babe Ruth, taken from the `Pitching` table.

yearID	teamID	G	GS	CG	W	L
1914	BOS	4	3	1	2	1
1915	BOS	32	28	16	18	8
1916	BOS	44	41	23	23	12
1917	BOS	41	38	35	24	13
1918	BOS	20	19	18	13	7
1919	BOS	17	15	12	9	5
1920	NYA	1	1	0	1	0
1921	NYA	2	1	0	2	0
1930	NYA	1	1	1	1	0
1933	NYA	1	1	1	1	0

statistics such as ERA (earned run average) and BAOpp (opponent's batting average).

Babe Ruth also provides a good illustration of the pitching statistics tables of Lahman's database since he had a great pitching record before becoming one of the greatest home run hitters in history. Table 1.4 displays statistics from the data table `Pitching` for the seasons in which Ruth was a pitcher. We see from the table that Ruth pitched in more than 40 games in 1916 and 1917 (by viewing column G), mostly as a starter (see GS), then appeared on the mound for half that many in the final two seasons for the Red Sox. When he moved to New York, he was only an occasional pitcher. Note that Ruth always was a winning pitcher as his wins (W) outnumbered his losses (L) for all pitching seasons, even when he returned to the pitching mound at the end of his career. He pitched one game both in 1930 and in 1933 (over ten years after he was a dominant pitcher for the Red Sox) and went the full nine innings (see variable CG) on each occasion.

1.2.6   The Fielding table

The `Fielding` table contains season-to-season fielding statistics for all players in major league history. For a given player, there will be a separate row for each fielding position. Outfielders positions are grouped together and labeled as OF for the older seasons, whereas for the more recent ones, they are conveniently distinguished as LF, CF, RF, for left fielders, center fielders, and right fielders, respectively. For a player in a position, the data tables give the count of games played (G), the count of games started (GS), the time played in the field expressed in terms of outs (InnOuts), the count of putouts (PO), assists (A), and errors (E).

To illustrate fielding data, Table 1.5 displays Babe Ruth's fielding statistics for his career. Only one row appears for each of the seasons between 1914 and 1917, as The Babe was exclusively employed as a pitcher. Later, as the Boston

TABLE 1.5

Fielding statistics for Babe Ruth, taken from the `Fielding` table. Columns featuring statistics relevant only to catchers are not reported.

playerID	yearID	stint	teamID	lgID	POS	G	GS	InnOuts	PO	A	E	DP
ruthba01	1914	1	BOS	AL	P	4	NA	NA	0	7	0	0
ruthba01	1915	1	BOS	AL	P	32	NA	NA	17	63	2	3
ruthba01	1916	1	BOS	AL	P	44	NA	NA	24	83	3	6
ruthba01	1917	1	BOS	AL	P	41	NA	NA	19	101	2	4
ruthba01	1918	1	BOS	AL	1B	13	NA	NA	130	6	5	8
ruthba01	1918	1	BOS	AL	OF	59	NA	NA	121	8	7	3
ruthba01	1918	1	BOS	AL	P	20	NA	NA	19	58	6	5
ruthba01	1919	1	BOS	AL	1B	5	NA	NA	35	4	1	4
ruthba01	1919	1	BOS	AL	OF	111	NA	NA	222	14	1	6
ruthba01	1919	1	BOS	AL	P	17	NA	NA	13	35	2	1
ruthba01	1920	1	NYA	AL	1B	2	NA	NA	10	0	1	1
ruthba01	1920	1	NYA	AL	OF	141	NA	NA	259	21	19	3
ruthba01	1920	1	NYA	AL	P	1	NA	NA	1	0	0	0
ruthba01	1921	1	NYA	AL	1B	2	NA	NA	8	0	0	0
ruthba01	1921	1	NYA	AL	OF	152	NA	NA	348	17	13	6
ruthba01	1921	1	NYA	AL	P	2	NA	NA	1	2	0	0
ruthba01	1922	1	NYA	AL	1B	1	NA	NA	0	0	0	0
ruthba01	1922	1	NYA	AL	OF	110	NA	NA	226	14	9	3
ruthba01	1923	1	NYA	AL	1B	4	NA	NA	41	1	1	2
ruthba01	1923	1	NYA	AL	OF	148	NA	NA	378	20	11	2
ruthba01	1924	1	NYA	AL	OF	152	NA	NA	340	18	14	4
ruthba01	1925	1	NYA	AL	OF	98	NA	NA	207	15	6	3
ruthba01	1926	1	NYA	AL	1B	2	NA	NA	10	0	0	2
ruthba01	1926	1	NYA	AL	OF	149	NA	NA	308	11	7	5
ruthba01	1927	1	NYA	AL	OF	151	NA	NA	328	14	13	4
ruthba01	1928	1	NYA	AL	OF	154	NA	NA	304	9	8	0
ruthba01	1929	1	NYA	AL	OF	133	NA	NA	240	5	4	2
ruthba01	1930	1	NYA	AL	OF	144	NA	NA	266	10	10	0
ruthba01	1930	1	NYA	AL	P	1	NA	NA	0	4	0	2
ruthba01	1931	1	NYA	AL	1B	1	NA	NA	5	0	0	0
ruthba01	1931	1	NYA	AL	OF	142	NA	NA	237	5	7	2
ruthba01	1932	1	NYA	AL	1B	1	NA	NA	3	0	0	0
ruthba01	1932	1	NYA	AL	OF	128	NA	NA	209	10	9	1
ruthba01	1933	1	NYA	AL	1B	1	NA	NA	6	0	1	0
ruthba01	1933	1	NYA	AL	OF	132	NA	NA	215	9	7	4
ruthba01	1933	1	NYA	AL	P	1	NA	NA	1	1	0	0
ruthba01	1934	1	NYA	AL	OF	111	NA	NA	197	3	8	0
ruthba01	1935	1	BSN	NL	OF	26	NA	NA	39	1	2	0

Red Sox took advantage of his powerful bat, there are three rows for 1918, one for each defensive position played by Ruth during this season.

Suppose one focuses on Ruth's fielding as an outfielder. One raw way of measuring his fielding range, proposed by Bill James in 1977 in his first *Baseball Abstract* (James 1980), is to sum his putouts (variable PO) and assists (variable A) and divide the sum by the games played (G). The values of this range factor' statistic for the seasons 1918 through 1935 were

```
[1] 2.19 2.13 1.99 2.40 2.18 2.69 2.36 2.27 2.14 2.26 2.03 1.84
[13] 1.92 1.70 1.71 1.70 1.80 1.54
```

Clearly, Ruth's range as an outfielder deteriorated toward the end of his career.

1.2.7   The Teams table

The Teams table contains seasonal data at the team level going back to 1871. A single row in this table includes the team's abbreviation (teamID), its final position in the standings (rank), its number of wins and losses (W and L), and whether the team won the World Series (WSWin), the League (LgWin), the Division (DivWin), or reached the post-season via the Wild Card (WCWin).

In addition, this table includes cumulative team offensive statistics such as counts of runs scored (R), hits (H), doubles (2B), walks (BB), strikeouts (SO), stolen bases (SB), and sacrifice flies (SF). Team defensive statistics include opponents runs scored (RA), earned runs allowed (ER), complete games (CG), shutouts (SHO), saves (SV), hits allowed (HA), home runs allowed (HRA), strikeouts by pitchers (SOA), and walks by pitchers (BBA). Team fielding statistics are included such as counts of errors (E), double plays (DP), and fielding percentage (FP). Last, this table includes the total home attendance (attendance) and the three-year park factors[3] for batters (BPF) and pitchers (PPF). Teams are identified, in this and other tables in the database, by a three-character code (teamID). The column name in the Teams table helps in recognizing clubs by their full name.

To illustrate the teams dataset, we extract the data for one of the greatest teams in baseball history, the 1927 New York Yankees.

```
  yearID lgID teamID franchID divID Rank   G Ghome   W   L
1   1927   AL    NYA      NYY  <NA>   1  155    77  110  44
  DivWin WCWin LgWin WSWin   R   AB    H X2B X3B HR  BB  SO SB
1   <NA>  <NA>    Y     Y  975 5347 1644 291 103 158 635 605 90
  CS HBP SF  RA  ER ERA CG SHO SV IPouts   HA HRA BBA SOA   E
1 64  NA NA 599 494 3.2 82  11 20   4167 1403  42 409 431 196
  DP     FP                name            park attendance BPF
```

---

[3]See Chapter 11 for an introduction to park factors.

```
1 123 0.969 New York Yankees Yankee Stadium I     1164015   98
  PPF teamIDBR teamIDlahman45 teamIDretro
1  94     NYY           NYA         NYA
```

We see the 1927 Yankees finished the season with a 110-44 record and won the World Series. The "Bronx Bombers" hit 158 home runs, stole 90 bases, and had a total home attendance of 1,164,015.

1.2.8 Baseball questions

The following questions can be answered with the Lahman database.

- [Q] **What is the average number of home runs per game recorded in each decade? Does the rate of strikeouts show any correlation with the rate of home runs?**

- [A] The number of home runs per game soared from 0.3 in baseball's first two decades to 0.8 in the 1920s. After the 1920s, the home run rate showed a steady increase up to 2.2 per game at the turn of the millennium. The first years of the current decade seem to reflect a decline in home run hitting as the rate has decreased to 1.9 HR per game. Strikeouts have steadily increased over the history of baseball—the number of strikeouts per game was 1 in the 1870s to 5.6 in the 1920s to 14.2 of the 2010s.

  – *Relevant data to obtain this answer is found in the* Teams *table.*

- [Q] **What effect has the introduction of the Designated Hitter (DH) in the American League had in the difference in run scoring between the American and National Leagues?**

- [A] The DH rule was instituted in 1973 only for the American League. Twice in the previous three years the National League teams had scored half a run more per game than the American League teams. From 1973 till the end of the decade run scoring was roughly equal. Since then, the American League has maintained an edge of about half a run per game.

  – *Relevant data to obtain this answer is found in the* Teams *table.*

- [Q] **How does the percentage of games completed by the starting pitcher from 2000 to 2010 compare to the percentage of games 100 years before?**

- [A] From 1900 to 1909 pitchers completed 79% of the games they started; from 2000 to 2010 it had dropped to 3.5%.

  – *Data for this answer can be found in the* Pitching *table.*

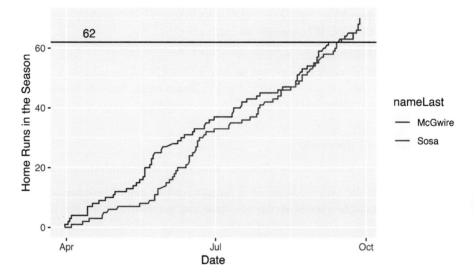

FIGURE 1.2
Home runs for Mark McGwire and Sammy Sosa during the 1998 race. The
horizontal line corresponds to a new season record of 62 home runs.

1.3   Retrosheet Game-by-Game Data

1.3.1   The 1998 McGwire and Sosa home run race

Another sacred Babe Ruth record was the 60 home runs recorded in the 1927
season. This record was eventually broken in 1961 by Roger Maris, after a
thrilling race with his teammate Mickey Mantle: the "M&M Brothers", as they
were often dubbed, ended the season with 61 and 54 home runs, respectively.
The new home run record lasted another 37 years. In 1998 two other players,
Mark McGwire of the St. Louis Cardinals and Sammy Sosa of the Chicago
Cubs, gave life to a new home run race, which is displayed in Figure 1.2. This
graph shows the cumulative home run count of each player as a function of
the day of the 1998 season.

From the figure, we see that for much of the season, McGwire was the only
man in the chase. Then Sosa caught fire and the two were very close in home
runs starting from mid-August. "Big Mac" first broke the record, hitting his
62nd home run on September 8. Then, on September 25, the two were tied at
66 apiece. Finally, McGwire managed to hit four more in the final days of the
season, while "Slammin' Sammy" remained at 66.

To produce the graph in Figure 1.2 and relive the 1998 season, one needs data
at a game-by-game level.

1.3.2 Retrosheet

Retrosheet is a volunteer organization, founded in 1989 by University of Delaware professor David Smith, that aims to collect play-by-play accounts of every game played in Major League Baseball history. Through the labor of love of many volunteers who have unearthed old newspaper accounts, scanned microfilms, and manually entered data into computers, the Retrosheet website[4] contains game-by-game summaries going back to the dawn of Major League Baseball in the 19th century. The Retrosheet site also has play-by-play data of most of the games played since the 1913 season and continues to add games for previous seasons. These data are introduced in Section 1.4.

1.3.3 Game logs

Retrosheet provides individual game data going back to 1871. A game log has details regarding when the game was played, how many spectators attended, the teams and the ballpark, and the score (both the final score and the inning by inning runs scored). In addition, the game log file includes teams' offensive and defensive statistics, starting players, managers, and umpire crews. There are missing observations for some game log variables for earlier baseball seasons.

Retrosheet provides a comprehensive *Guide to Retrosheet Game Logs*[5] document that gives details of all 161 fields compiled for each game. Readers are encouraged to peruse the guide to fully understand the contents of the files. Details on the relevant data fields will be described when they are used in later chapters.

1.3.4 Obtaining the game logs from Retrosheet

Game log files can be found at https://www.retrosheet.org/gamelogs/index. html. A zip file is provided for each season, starting from 1871, and can be downloaded in a folder of choice by clicking on the relevant year. When one extracts the zip file, one obtains a plain text file (`.txt` extension) where fields are separated by commas. Section 11.4 provides an R function for downloading and parsing game log files.

1.3.5 Game log example

On September 9, 1995, Cal Ripken, Jr. of the Baltimore Orioles surpassed the seemingly unbeatable consecutive games record of 2130 belonging to the late Lou Gehrig. One can learn more about this historic game by exploring the game log files for the 1995 season. Table 1.6 contains a subset of the copious information available for this particular game between Baltimore and

---

[4]https://www.retrosheet.org
[5]https://www.retrosheet.org/gamelogs/glfields.txt

TABLE 1.6

Excerpt of data available in the Retrosheet game logs. Sample from the Cal Ripken's Iron Man game (Sept. 6, 1995).

Variable	Value
date	19950906
dayofweek	Wed
visitorteam	CAL
hometeam	BAL
visitorrunsscored	2
homerunsscore	4
daynight	N
parkid	BAL12
attendance	46272
duration	215
visitorlinescore	100000010
homelinescore	10020010x
homeab	34
homeh	9
homehr	4
homerbi	4
homebb	1
homek	8
homegdp	0
homelob	7
homepo	27
homea	8
homee	0
umpirehname	Larry Barnett
umpire1bname	Greg Kosc
umpire2bname	Dan Morrison
umpire3bname	Al Clark
visitormanagername	Marcel Lacheman
homemanagername	Phil Regan
homestartingpitchername	Mike Mussina
homebatting1name	Brady Anderson
homebatting1position	8
homebatting2name	Manny Alexander
homebatting2position	4
homebatting3name	Rafael Palmeiro
homebatting3position	3
homebatting4name	Bobby Bonilla
homebatting4position	9
homebatting5name	Cal Ripken
homebatting5position	6

(*Continued on next page*)

TABLE 1.6
(*Continued*).

Variable	Value
homebatting6name	Harold Baines
homebatting6position	10
homebatting7name	Chris Hoiles
homebatting7position	2
homebatting8name	Jeff Huson
homebatting8position	5
homebatting9name	Mark Smith
homebatting9position	7

California. These data are taken from a single line in the `gl1995.txt` file available at https://www.retrosheet.org/gamelogs/index.html. This table displays team statistics[6] as well as the players' identities and fielding positions for the home team; similar statistics and player information are available for the visiting team.

What does one learn from this game log information displayed in Table 1.6? This game took place on a Wednesday night in front of 46,272 people in Baltimore (the `hometeam = BAL` indicates the Orioles were the home team). The game lasted over three and a half hours (`duration = 215` minutes), thanks in part to the standing ovation Ripken got at the end of the fifth inning, when the game became official. (The standing ovation information is not available in this file.) Baltimore defeated California 4-2; since we observe `homehr = 4`, we observe that all of Baltimore's runs this game were due to four home runs with the bases empty. The Orioles infield in this game included Rafael Palmeiro at first base, Manny Alexander at second base, Ripken at shortstop, and Jeff Huson at third base.

1.3.6 Baseball questions

Here are some typical questions one can answer with the Retrosheet game logs files.

- [Q] **In which months are home runs more likely to occur? What about ballparks?**

- [A] Since 1980, July has been the month with the most home runs per game (1.97), while September has had the lowest frequency (1.84). In the same time frame, 2.71 home runs per game have been hit in Coors Field

---

[6]Some other team statistics omitted in Table 1.6—such as Stolen Bases and Caught Stealing—are reported in game log files.

(home of the Colorado Rockies, and 1.14 in the Astrodome (the former home of the Houston Astros.

- [Q] **Do runs happen more frequently when some umpires are behind the plate? What is the difference between the most pitcher-friendly and the most hitter-friendly umpires?**

- [A] Among umpires with more than 400 games called since 1980, teams scored the highest number of runs (10.0 per game combined) when Chuck Meriwether was behind the plate and the lowest (7.8) when Doug Harvey was in charge.

- [Q] **How many extra people attend ballgames during the weekend? What's the average attendance by day of the week?**

- [A] Close to 33,000 people attend games played on Saturdays (data from 1980 to 2011) and 31,000 on Sundays. The average goes down to 29,000 on Fridays, 25,000 on Thursdays and Mondays, and 24,000 on Tuesdays and Wednesdays.

1.4 Retrosheet Play-by-Play Data

1.4.1 Event files

Retrosheet has collected data to an even finer detail for most games played since 1913. For those seasons, play-by-play accounts are available at https://www.retrosheet.org/game.htm. These "event files" (as these play-by-play files are named) contain information on every single event happening on the field during a game. For each play, information is reported on the situation (inning, team batting, number of outs, presence of runners on base), the players on the field, the sequence of pitches thrown, and details on the play itself. For example, the file indicates whether a hit occurred, and if a ball in play is a ground ball, the file gives the defender that fielded the ball.

Event files come in a format expressly devised for them. Retrosheet gives detailed instruction on how to use the files[7] and a step-by-step guide[8], plus the software to parse the files.[9] However, the process of rendering the files in a format suitable for use in R (or other statistical programs) is not straightforward without use of additional tools. Thus, in Appendix A, we present R code that implements the full process of downloading, extracting, and parsing these data. We also provide sample code to create the datasets used in this book.

---

[7]How to use Our Event Files: https://www.retrosheet.org/datause.txt
[8]Step-by-Step Example: https://www.retrosheet.org/stepex.txt
[9]Software Tools: https://www.retrosheet.org/tools.htm

TABLE 1.7

Excerpt of information available in Retrosheet event files. Sample from Jeter's "Flip Play" (Oct. 13, 2001).

Variable	Value
GAME_ID	OAK200110130
YEAR_ID	2001
AWAY_TEAM_ID	NYA
INN_CT	7
BAT_HOME_ID	1
OUTS_CT	2
BALLS_CT	2
STRIKES_CT	2
PITCH_SEQ_TX	CSBBFX
AWAY_SCORE_CT	1
HOME_SCORE_CT	0
BAT_ID	longt002
BAT_HAND_CD	L
PIT_ID	mussm001
PIT_HAND_CD	R
POS2_FLD_ID	posaj001
POS3_FLD_ID	martt002
POS4_FLD_ID	soria001
POS5_FLD_ID	bross001
POS6_FLD_ID	jeted001
POS7_FLD_ID	knobc001
POS8_FLD_ID	willb002
POS9_FLD_ID	spens001
BASE1_RUN_ID	giamj002
BASE2_RUN_ID	NA
BASE3_RUN_ID	NA
EVENT_TX	D9/9S.1XH(962)
BAT_FLD_CD	7
BAT_LINEUP_ID	7

1.4.2   Event example

Just as we use a historical game for the purpose of showing the contents of Retrosheet game logs, we use a famous fielding play to illustrate the Retrosheet event files. This play is represented as a single line in an event file shown in Table 1.7.

The play took place in a game played in Oakland on October 13, 2001, as can be inferred from the value of the GAME_ID variable. This game was Game 3 of the American League Division Series featuring the hometown Athletics against the New York Yankees (AWAY_TEAM_ID = NYA). The play occurred in the seventh inning with the home team batting (variables INN_CT and BAT_ID_ID). There were two outs (variable OUTS_CT) and the A's were leading 1-0 (variables AWAY_SCORE_CT and HOME_SCORE_CT). Right-handed Mike Mussina (variables PIT_ID and PIT_HAND_CD) was on the mound for the Yankees, facing left-handed batter Terrence Long (variables BAT_ID and BAT_HAND_CD) with Jeremy Giambi standing on first base (variable BASE1_RUN_ID). The BAT_FLD_CD = 7 and BAT_LINEUP_ID = 7 fields inform us that Giambi's defensive position was left field (position 7 corresponding to left field) and he was batting 7th in the lineup. The variables POS2_FLD_ID through POS9_FLD_ID report the full defensive lineup for the Yankees.

The seemingly inscrutable characters appearing in the PITCH_SEQ_TX and EVENT_TX variables depict what happened during that particular at bat. From looking at the pitch sequence variable PITCH_SEQ_TX, one sees that Mussina quickly went ahead in the count as Long let a strike go by and swung and missed at another pitch (CS). Then Mussina followed with consecutive balls (BB) and Long battled with a foul ball (F) before putting the ball in play (X). The variable EVENT_TX gives the results of the play. Long's hit resulted in a double, collected by the Yankees right fielder (D9 in EVENT_TX) in short right (9S). The runner on first was thrown out on his way to home (1XH) by a throw from right fielder Shane Spencer, relayed by shortstop Derek Jeter to catcher Jorge Posada (962).

Once the event files are properly processed, many more fields are available than the ones presented in Table 1.7. However these additional fields are, for the most part, derived from what is in the table. For example, one additional field indicates whether the at-bat resulted in a base hit, one field will identify the fielder who collected the ball, and four fields will indicate where each runner (and the batter) stood at the end of the play—all of this can be inferred by the EVENT_TX field.

This play-by-play information is available for most games going back to 1913; thus it is possible to recreate what happened on the field in the past century. For this particular play, the Retrosheet event files cannot tell us all of the interesting details. Derek Jeter came out of nowhere to cut off Spencer's throw and flipped it backhand to Posada in time to nail Giambi at home, on what has become known as "The Flip Play".[10]

1.4.3 Baseball questions

Below are some questions that can be explored with the Retrosheet event files. These specific questions are about how batters perform in particular situations in the pitch count and with runners on base.

---

[10]In 2002, *Baseball Weekly* recognized "The Flip Play" as one of the ten most amazing fielding plays of all time.

- [Q] **During the McGwire/Sosa home run race, which player was more successful at hitting homers with men on base?**

- [A] Mark McGwire hit 37 home runs in 313 plate appearances with runners on base, while Sammy Sosa hit 29 in 367. Once walks (both intentional and unintentional) and hit by pitches are removed, the number of opportunities become 223 for McGwire and 317 for Sosa.

- [Q] **How many intentional walks in unusual situations (e.g., empty bases or bases loaded) was Barry Bonds issued in his 73 HR campaign?**

- [A] During his record 2001 season, Barry Bonds was passed intentionally only 35 times. Of those free passes, one came with a runner on first and two with runners on first and second. When he was awarded 120 intentional walks in 2004, 19 came with nobody on, 11 with a runner on first, and 3 with runners on first and third. He was once walked intentionally with the bases loaded in 1998.

- [Q] **What is the Major league batting average when the ball/strike count is 0-2? What about on 2-0?**

- [A] In 2011, hitters compiled a .253 batting average on plate appearances where they fell behind 0-2. Conversely they hit .479 after going ahead 2-0.

1.5   Pitch-by-Pitch Data

1.5.1   MLBAM Gameday and PITCHf/x

Erstwhile Miami Marlins rightfielder Giancarlo Stanton emerged in 2017 as an elite slugger by blasting a league leading 59 balls out of the park. Figure 1.3 shows the location and the type of the 59 pitches Stanton sent into the stands.

Since 2005, baseball fans have had the opportunity to follow, pitch-by-pitch, the games played by their favorite team on the Web thanks to the Major League Baseball Advanced Media (MLBAM) Gameday application featured on the MLB.com website. For a couple of years, fans would only know the outcome of each pitch (whether it was a ball, a called strike, a swinging strike, and so on). Starting from an October 2006 game played at the Metrodome in Minneapolis, a wealth of detail began to appear for each pitch tracked on Gameday. Data on the release point, the pitch speed, and its full trajectory, were available for about one-third of the games played in 2007. Starting from the 2008 season, nearly every MLB pitch flight has been recorded by the PITCHf/x system. However, since the second edition of this book, PITCHf/x has been superseded by Statcast, a data source which is detailed in Section 1.6.1. What we provide here serves two purposes: first, it provides historical continuity; and second,

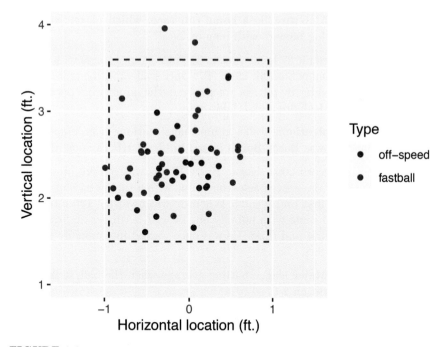

FIGURE 1.3
Pitch type and location for Giancarlo Stanton's 59 home runs of the 2017 season.

most of the information recorded by PITCHf/x appears in a similar format in the Statcast data, so the content is still relevant. Unfortunately, we are not aware of a currently usable method for obtaining pitch-by-pitch data for the pre-Statcast period. Please see Appendix B for a fuller discussion of the rise and fall of PITCHf/x.

1.5.2   PITCHf/x Example

On April 21, 2012, Phil Humber became the 21st pitcher in Major League Baseball history to throw a perfect game by retiring all the 27 batters he faced. PITCHf/x captured his final pitch (like it did for nearly every other pitch thrown in MLB ballparks from 2008 to 2015), providing the data shown in Table 1.8. The outcome of the pitch (variable des) is recorded by a stringer (a human being), while most of the remaining information is either captured by the Sportvision camera system or calculated from the captured data.

Each pitch is assigned an identifier (sv_id), that is actually a time stamp: Humber's final pitch was recorded on April 21, 2012, at 15:25:37. The key information Sportvision obtains through its camera system is recorded in lines 11 through 19 of the table. Those nine parameters give the position (variables x0, y0, z0), velocity (variables vx0, vy0, vz0), and acceleration (variables ax,

TABLE 1.8

Excerpt of information available from PITCHf/x. Sample from the final pitch of Phil Humber's perfect game (Apr. 21, 2012).

Variable	Value
des	Swinging Strike (Blocked)
sv_id	120421_152537
start_speed	85.3
end_speed	79.1
sz_top	3.73
sz_bot	1.74
pfx_x	0.31
pfx_z	1.81
px	2.211
pz	1.17
x0	−1.58
y0	50.0
z0	5.746
vx0	9.228
vy0	−124.71
vz0	−5.311
ax	0.483
ay	25.576
az	−29.254
break_y	23.8
break_angle	−4.1
break_length	7.8
pitch_type	SL
spin_dir	170.609
spin_rate	344.307

ay, az) components of the pitch at release point. With these nine parameters the full trajectory of the pitch from release to home plate can be estimated. (In fact, Sportvision actually estimates the parameters somewhere in the middle of the ball's flight, then derives the parameters at release point.)

While the nine parameters just mentioned are sufficient for learning about the trajectory of the pitch, they are difficult to understand by casual fans who follow the game on MLBAM Gameday. Other more descriptive quantities are calculated starting from those nine parameters. The one measure familiar to baseball fans is the pitch speed at release, which for Humber's final pitch is calculated at 85.3 mph (variable `start_speed`). PITCHf/x also provides the speed of the ball as it crosses the plate, 79.1 mph in this case (variable `end_speed`). Another two important values are the variables `px` and `pz`; they represent the horizontal and vertical location of the pitches, respectively, and can be combined with the batter's strike zone upper and lower limits (`sz_top` and `sz_bot`) to infer whether the pitch crossed the strike zone.

Let's focus on the location of this particular pitch. The horizontal reference point is the middle of the plate, with positive values indicating pitches passing on the right side of it from the umpire's viewpoint. In this case the ball crossed the plate 2.21 feet on the right of its midpoint. Since the plate is 17 inches wide, it was way out of the strike zone. The pitch was also too low to be a strike, as the vertical point at which it crossed the plate is listed at 1.17 feet, while the hitter's lower limit of the strike zone is 1.74.[11] Luckily for Humber (since otherwise a walk would have ruined the perfect game), the home plate umpire controversially declared that Brendan Ryan had swung the bat for strike three.

Other interesting quantities about a pitch are available with PITCHf/x, including the horizontal and vertical movement (variables `pfx_x` and `pfx_z`) of the pitch trajectory, the spin direction, and its spin rate (variables `spin_dir` and `spin_rate`). MLBAM has devised a complex algorithm that processes the information captured by Sportvision and marks the pitch with a label familiar to baseball fans. In this case the algorithm recognizes the pitch as a slider (variable `pitch_type`).

1.5.3   Baseball questions

Below are questions you can answer with PITCHf/x data. The data can be used to address specific questions about pitch type, speed of the pitch, and play outcomes on specific pitches.

- [Q] **Who are the hitters who see the lowest and the highest percentage of fastballs?**

- [A] From 2008 to 2011, pitchers have thrown fastballs 35% of the time when Ryan Howard was at the plate, 56% of the time when facing David Eckstein.

- [Q] **Who is the fastest pitcher in baseball currently?**

- [A] Nine of the fastest ten pitches recorded by PITCHf/x from 2008 to 2011 have been thrown by Aroldis Chapman, the highest figure being a 105.1 mph pitch thrown on September 24, 2010 in San Diego. Neftali Feliz is the other pitcher making the top ten list with a 103.4 fastball delivered in Kansas City on September 1, 2010.

- [Q] **What are the chances of a successful steal when the pitcher throws a fastball compared to when a curve is delivered?**

---

[11]The batter's strike zone boundaries are recorded by the human stringer at the beginning of the at-bat, and thus are less precise than the pitch location coordinates recorded by the advanced system.

- [A] From 2008 to 2010, baserunners were successful 73% of the times at stealing second base on a fastball. The success rate increases to 85% when the pitch is a curveball.

1.6 Player Movement and Off-the-Bat Data

1.6.1 Statcast

Statcast is a new technology that tracks movements of all baseball players and the baseball during a game. It was introduced to all MLB stadiums in 2015, and all teams have access to the large amount of data collected. Some of the Statcast data for pitchers and hitters is currently publicly available via the Baseball Savant website. Also, these data are used during a baseball broadcast for entertainment purposes. For example, a baseball announcer will provide the launch angle, exit velocity, and distance traveled for a home run. The broadcast will give the distance that an outfielder moves toward a batted ball and the speed of a baserunner from home to first base.

1.6.2 Baseball Savant data

To illustrate the new information available from Statcast, consider data on one of the hardest hit home runs during the 2018 season. In a game between the New York Yankees at the Toronto Blue Jays on June 6, 2018, Giancarlo Stanton of the Yankees hit a home run in the top of the 13th inning. The **baseballr** package provides a function to download data on each pitch from the Baseball Savant website, and Table 1.9 displays a number of variables for this specific home run.

The batter id and pitcher id are respectively 519317 and 607352 which correspond to Giancarlo Stanton and Joe Biagini. The game situation variables are `inning`, `inning_topbot`, `outs_when_up`, `home_score`, `away_score`, `balls`, and `strikes`. We learn that this home run was hit in the top of the 13th inning when the Yankees were leading 2-0, there were two outs, and the count was 1-0.

The variables `if_fielding_alignment` and `of_fielding_alignment` relate to the positioning of the Toronto fielders for this particular pitch. Both variables are "Standard", which indicate that there was no special shifting of the infielders or outfielders for Stanton for this particular at-bat.

Other variables give characteristics of the pitch. The `pitch_type` and `pitch_name` variables indicate that the pitch was a change-up thrown at a `release_speed` of 85.4 mph. The `plate_x` and `plate_y` variables give the location of the pitch—these values indicate that the pitch was located in the middle of the zone.

TABLE 1.9
Statcast data on home run hit by Giancarlo Stanton on game on June 6, 2018.

Variable	Value
pitch_type	CH
game_date	2018-06-06
release_speed	85.4
batter	519317
pitcher	607352
events	home_run
des	Giancarlo Stanton homers (14) on a line drive to left field.
home_team	TOR
away_team	NYY
balls	1
strikes	0
game_year	2018
plate_x	−0.2197
plate_z	2.4995
outs_when_up	2
inning	13
inning_topbot	Top
hc_x	20.36
hc_y	56.47
hit_distance_sc	416
launch_speed	119.3
launch_angle	15.044
pitch_name	Changeup
home_score	0
away_score	2
if_fielding_alignment	Standard
of_fielding_alignment	Standard
barrel	1

This dataset also includes characteristics of the batted ball. The `launch_speed` and `launch_angle` variables tell us that the ball came off of the bat at a speed of 119.3 mph at a launch angle of 15.044 degrees. It is notable that this particular batted ball was a line drive—it must have been hit hard for a home run. The `hit_distance_sc` variable indicates that the ball was hit a distance of 416 feet. The `hc_x` and `hc_y` variables tell us about the direction (more specifically, the spray angle of this home run.

More details about the Statcast data and how to use it appear in Appendix C.

1.6.3 Baseball questions

The following questions can be addressed with Statcast data.

- [Q] **What is a typical launch speed and launch angle of a home run?**
- [A] Using 2017 season data, the median launch speed of a home run was 103 mph and the median launch angle was 27.8 degrees.
- [Q] **How frequently do MLB teams employ infield shifts?**
- [A] Using data from the first part of the 2018 season, teams employed an infield shift or some strategic infield defense for 26 percent of the batters.
- [Q] **Is an infield shift effective in defensing ground balls?**
- [A] Using data from the 2018 season, the batting average on ground balls with a standard infield defense was 0.281, and with an infield shift the batting average on ground balls dropped to 0.231.

1.7 Data Used in this Book

We use data from all four of the sources above in various places in this book. Generally, data that comes from the Lahman database will be accessed directly from the R package **Lahman**. Small data sets are available through the **abdwr3edata** package, the full source of which is located on GitHub. The Retrosheet and Statcast data is large enough that it is generally not included in the GitHub repositories. In order to reproduce the results in this book using those datasets, you will need to download those data on your own. Instructions for doing so appear in Section A.1.3 (for Retrosheet) and Section 12.2 (for Statcast).

1.8 Summary

When choosing among the four main sources of baseball data (Lahman, Retrosheet, PITCHf/x, and Statcast, one always has to consider the trade-off between the level of detail and the seasons covered by the source. With Lahman's database, for example, one can explore the evolution of the game since its beginnings back into the 19th century. However, only the basic season count statistics are available from this source. For example, simple information such as Babe Ruth's batting splits by pitcher's handedness cannot be retrieved from Lahman's files.

Retrosheet is steadily adding past seasons to its play-by-play database, allowing researchers to perform studies to validate or reject common beliefs about players of the past decades. During the years, for example, analysis of play-by-play data has confirmed the huge defensive value of players like Brooks Robinson and Mark Belanger, and has substantiated the greatness of Roberto Clemente's throwing arm.

PITCHf/x was available from 2008–2015 and, unlike with Retrosheet, there is no way to compile data for games of the past. This means we will never be able to compare the velocity of Aroldis Chapman's fastball to that of Nolan Ryan or Bob Feller. However, studies performed since its inception have provided an enhanced understanding of the game, enabling researchers to explore issues like pitch sequencing, batter discipline, pitcher fatigue, catcher framing (see Chapter 7) and the catcher's ability to block bad pitches.

Statcast represents the newest wave of baseball data. Since it includes detailed information on the positioning and movement of players, it has been useful in evaluating the effectiveness of a fielder in reaching a batted ball and understanding the performance of runners on the bases. Also, Statcast has helped us understand the relationship between the batted ball variables (launch angle, exit velocity, and spray angle and base hits and outs.

1.9   Further Reading

Schwarz (2004) provides a detailed history of baseball statistics. Adler (2006) explains how to obtain baseball data from several sources, including Lahman's database, Retrosheet, and MLBAM Gameday and how to analyze them using diverse tools, from Microsoft Excel to R, MySQL, and PERL. Fast (2010) introduces the PITCHf/x system to the uninitiated. The *PITCHf/x, HITf/x, FIELDf/x* section of *The Physics of Baseball* website (http://baseball.physics .illinois.edu/) features material on the subject of pitch tracking data.

1.10   Exercises

1. Which Datafile?

This chapter has given an overview of the Lahman database, the Retrosheet game logs, the Retrosheet play-by-play files, the PITCHf/x database, and the Statcast database. Describe the relevant data among these four databases that can be used to answer the following baseball questions.

a. How has the rate of walks (per team for nine innings) changed over the history of baseball?

b. What fraction of baseball games in 1968 were shutouts? Compare this fraction with the fraction of shutouts in the 2012 baseball season.

c. What percentage of first pitches are strikes? If the count is 2-0, what fraction of the pitches are strikes?

d. Which players are most likely to hit groundballs? Of these players, compare the speeds at which these groundballs are hit.

e. Is it easier to steal second base or third base? (Compare the fraction of successful steals of second base with the fraction of successful steals of third base.)

2. Lahman Pitching Data

From the pitching data file from the Lahman database, the following information is collected about Bob Gibson's famous 1968 season.

```
  playerID yearID stint teamID lgID  W L  G GS CG SHO SV
1 gibsobo01  1968    1    SLN   NL 22 9 34 34 28  13  0
  IPouts   H ER HR BB  SO BAOpp  ERA IBB WP HBP BK  BFP GF  R
1    914 198 38 11 62 268  0.18 1.12   6  4   7  0 1161  0 49
  SH SF GIDP
1 NA NA   NA
```

a. Gibson started 34 games for the Cardinals in 1968. What fraction of these games were completed by Gibson?

b. What was Gibson's ratio of strikeouts to walks this season?

c. One can compute Gibson's innings pitched by dividing IPouts by three. How many innings did Gibson pitch this season?

d. A modern measure of pitching effectiveness is WHIP, the average number of hits and walks allowed per inning. What was Gibson's WHIP for the 1968 season?

3. Retrosheet Game Log

Jim Bunning pitched a perfect game on Father's Day on June 21, 1964. Some details about this particular game can be found from the Retrosheet game logs.

```
      Date DoubleHeader DayOfWeek VisitingTeam
1 19640621            1       Sun          PHI
  VisitingTeamLeague VisitingTeamGameNumber HomeTeam
1                 NL                     60      NYN
  HomeTeamLeague HomeTeamGameNumber VisitorRunsScored
1             NL                 67                 6
  HomeRunsScore LengthInOuts DayNight CompletionInfo
1             0           54        D             NA
  ForfeitInfo ProtestInfo ParkID Attendance Duration
```

1	NA	NYC17	0	139

	VisitorLineScore	HomeLineScore	VisitorAB	VisitorH	VisitorD
1	110004000	000000000	32	8	2

	VisitorT	VisitorHR	VisitorRBI	VisitorSH	VisitorSF	VisitorHBP
1	0	1	6	2	0	0

	VisitorBB	VisitorIBB	VisitorK	VisitorSB	VisitorCS	VisitorGDP
1	4	0	6	0	1	0

	VisitorCI	VisitorLOB
1	0	5

a. What was the time in hours and minutes of this particular game?
b. Why is the attendance value in this record equal to zero?
c. How many extra base hits did the Phillies have in this game? (We know that the Mets had no extra base hits this game.)
d. What was the Phillies' on-base percentage in this game?

4. Retrosheet Play-by-Play Record

One of the famous plays in Philadelphia Phillies history is second baseman Mickey Morandini's unassisted triple play against the Pirates on September 20, 1992. The following records from the Retrosheet play-by-play database describe this half-inning. The variables indicate the half-inning (variables INN_CT and HOME_ID), the current score (variables AWAY_SCORE_CT and HOME_SCORE_CT), the identities of the pitcher and batter (variables BAT_ID and PIT_ID), the pitch sequence (variable PITCH_SEQ), the play event description (variable EVENT_TEX), and the runners on base (variables BASE1_RUN and BASE2_ID).

	bat_home_id	away_score_ct	home_score_ct	bat_id	pit_id
1	1	1	1	vansa001	schic002
2	1	1	1	bondb001	schic002
3	1	1	1	kingj001	schic002

	pitch_seq_tx	event_cd	event_tx	base1_run_id
1	CBBBX	20	S9/L9M	
2	C1BX	20	S/G56.1-2	vansa001
3	BLLBBX	2	4(B)4(2)4(1)/LTP/L4M	bondb001

	base2_run_id
1	
2	
3	vansa001

Based on the records, write a short paragraph that describes the play-by-play events of this particular inning.

5. PITCHf/x Record of Several Pitches

R.A. Dickey was one of the few current pitchers in recent years who threw a knuckleball. The following gives some PITCHf/x variables for the first knuckleball and the first fastball that Dickey threw for a game against the

Kansas City Royals on April 13, 2013.

start_speed	end_speed	pfx_x	pfx_z	px	pz	sz_bot	sz_top
73	66.3	-0.64	-7.58	-0.047	2.475	1.5	3.35

start_speed	end_speed	pfx_x	pfx_z	px	pz	sz_bot	sz_top
81.2	75.4	-4.99	-7.67	-1.99	2.963	1.5	3.43

Describe the differences between the knuckleball and the fastball in terms of pitch speed, movement (horizontal and vertical directions), and location in the strike zone. Based on this data, why is a knuckleball so difficult for a batter to make contact?

2

Introduction to R

2.1 Introduction

In this chapter, we provide a general introduction to the R statistical computing environment. We describe the process of installing R and the program RStudio that provides an attractive interface to the R system. We use pitching data from the legendary Warren Spahn to motivate manipulations with vectors, a basic data structure. We describe different data types such as characters, factors, and lists, and different "containers" for holding these different data types. We discuss the process of executing collections of R commands by means of scripts and functions, and describe methods for importing and exporting datasets from R. A fundamental data structure in R is a data frame and we introduce defining a data frame, performing manipulations, merging data frames, and performing operations on a data frame split by values of a variable. We conclude the chapter by describing how to install and load R packages and how one gets help using resources from the R system and the RStudio interface.

2.2 Installing R and RStudio

The R system is available for download from The Comprehensive R Archive Network (CRAN) at https://www.r-project.org. R is available for Linux, Windows, and Macintosh operating systems; all of the commands described in this book will work in any of these environments.

One can use R through the standard *graphical user interface* by launching the R application. Recently, several new integrated developmental environments have been created for R, and we will demonstrate the RStudio environment (RStudio Team 2018) available from (https://www.rstudio.com). One must first install R, then the RStudio application, and then launches the RStudio application. All interaction with R occurs through RStudio.

DOI: 10.1201/9781032668239-2

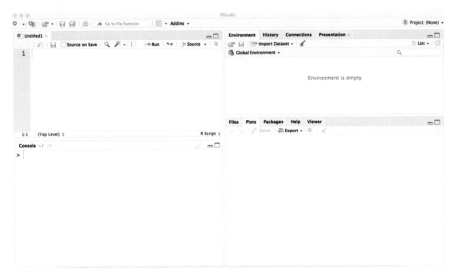

FIGURE 2.1
The opening screen of the RStudio interface to R.

The RStudio opening screen is displayed in Figure 2.1. The screen is divided into four windows. One can type commands directly and see output in the lower-left Console window. Moving clockwise, the top-left window is a blank file where one can write and execute R scripts or groups of instructions. The top-right window shows names of objects such as vectors and data frames created in an R session. By clicking on the History tab, one can see a record of all commands entered during the current R session. Last, any plots are displayed in the lower-right window. By clicking on the Files tab, one can see a list of files stored in the current *working directory*. (This is the file directory where R will expect to read files, and where any output, such as data files and graphs, will be stored.) The Packages tab lists all of the R packages currently installed in the system and the Help tab will display documentation for R functions and datasets.

2.3 The Tidyverse

R is a modular system—functionality can be added by installing and loading *packages* (see Section 2.9). The R language, along with some of the core packages (e.g., **base**, **stats**, and **graphics**) are developed by the R Core Development Team. However, packages can be written by anyone and can provide a wide variety of functionality. Recently, tremendous coordinated effort by many contributors has led to the development of the *tidyverse*. The tidyverse is a collection of packages intentionally designed for interoperability, centered around the philosophy of tidy data, articulated most notably by Posit Chief Scientist Hadley Wickham (Wickham 2014). The **tidyverse** package itself does little more than load a collection of other packages that adhere to this philosophy.

A major undertaking of the second edition of this book was to bring all of the code into **tidyverse**-compliance.

```
library(tidyverse)
```

The central notion of tidy data is that rows in a data frame should correspond to the same observational unit, and that columns should represent variables about those observational units. This means that a tidy data frame would not contain a row that totals the other rows, labels for the rows stored as row names instead of variables, or two columns that contain the same type of information about the same observational unit.

Packages in the tidyverse are `data.frame`-centric (see Section 2.4): their functions mostly take a `data.frame` as the first input, do something to it, and then return (a modified version of) it. Other data structures like matrices, vectors, and lists are less commonly used in the tidyverse. A `tibble` is like a `data.frame`, but may include some additional functionality. In this book, we prefer `tibbles` to `data.frames` whenever it is convenient.

There are many packages in the **tidyverse**, but a few warrant explicit introduction.

2.3.1   dplyr

The **dplyr** package provides comprehensive tools for data manipulation (or data wrangling). The five main "verbs" include:

- `select()`: choose a subset of the columns
- `filter()`: choose a subset of the rows based on logical criteria
- `arrange()`: sort the rows based on the values of columns
- `mutate()`: add or modify the definition of a column
- `summarize()`: collapse a data frame down to a single row (per group) by aggregating vectors into single values. Often used in conjunction with `group_by()`.

These five functions, along with functions that allow you to merge two data frames together by matching corresponding values (e.g., `inner_join()`, `left_join()`, etc.) can be combined to produce the functionality equivalent to a SQL SELECT query (see Chapter 11). A vast array of data analytic operations can be reduced to combinations of these functions (along with a few other verbs provided by **dplyr**, like `rename()`, `count()`, `bind_rows()`, `pull()`, etc.).

2.3.2 The pipe

A major component of the tidyverse design is the use of the *pipe operator*: |>. The pipe operator allows one to create pipelines of functions that are easier to read than their nested counterparts. The pipe takes what comes before it and injects it into the function that comes after it as the first argument. Thus, the following lines of code are equivalent.

```
outer_function(inner_function(x), y)

x |>
  inner_function() |>
  outer_function(y)
```

We find that the latter code chunk is easier to read, scales better to many successive operations (i.e., pipelines), and keeps arguments closer to the functions to which they belong. Since many functions in the tidyverse take a `tibble` as their first argument and return a `tibble`, they are (by design) "pipeable".

Since version 4.1.0, R has included the native pipe operator: |>. This operator is used extensively in this book, and replaces the legacy pipe operator (%>%) that was provided by the **magrittr** package and was used throughout the second edition of this book.

2.3.3 ggplot2

ggplot2 is the graphics system for the tidyverse. It is an implementation of *The Grammar of Graphics* (Wilkinson 2006), and provides a consistent syntax for building data graphics incrementally in layers. Please note that for historical and technical reasons, the plus operator + (rather than the pipe operator) is used to combine elements in **ggplot2**. We describe **ggplot2** in detail in Chapter 3.

2.3.4 Other packages

In addition to the aforementioned **dplyr**, **ggplot2**, and **tibble** packages, loading the **tidyverse** package also loads several other packages. These include **tidyr** for additional data manipulation operations, **readr** for data import (see Section 2.8), **purrr** for iteration (see Section 2.10), **stringr** for working with text (see Section 2.6.1), **lubridate** for working with dates, and **forcats** for working with factors (see Section 2.6.2). Other packages—like **broom**—are not loaded automatically, but are part of the larger tidyverse.

2.3.5 Package for this book

As mentioned earlier, the R package **abdwr3edata** contains all small datafiles and R scripts described in this book. One can install this package by use of `install_github()` from the **remotes** package:

```
remotes::install_github("beanumber/abdwr3edata")
```

Then the **abdwr3edata** package can be loaded into R by use of the `library()` function.

```
library(abdwr3edata)
```

> **!** Important
>
> Installation of the **abdwr3edata** package needs to be done only once, but the package should be loaded in *each new R session* that uses these datasets.

2.4 Data Frames

2.4.1 Career of Warren Spahn

One of the authors collected the 1965 Warren Spahn baseball card. The back of Spahn's baseball card displays many of the standard pitching statistics for the seasons preceding Spahn's final 1965 season. We use data from Spahn's season statistics to illustrate some basic components of the R system.

2.4.2 Introduction

A `data.frame` is a rectangular table of data, where rows of the table correspond to different individuals or seasons, and columns of the table correspond to different variables collected on the individuals. Data variables can be numeric (like a batting average or a winning percentage), integer (like the count of home runs or number of wins), a factor (a categorical variable such as the player's team), or other types.

We can display portions of a data frame using the square bracket notation. For example, if we wish to display the first five rows and the first four variables (columns) of a data frame x, we type `x[1 : 5, 1 : 4]`. Alternatively, the functions `slice()` and `select()` in the **dplyr** package can be used to select specific rows and columns. For example, the following code displays the first three rows and columns 1 though 10 of the `spahn` data frame.

```
library(abdwr3edata)
spahn |>
  slice(1:3) |>
  select(1:10)
```

```
# A tibble: 3 x 10
   Year   Age Tm    Lg        W     L   W.L   ERA     G    GS
  <dbl> <dbl> <chr> <chr> <dbl> <dbl> <dbl> <dbl> <dbl> <dbl>
1  1942    21 BSN   NL        0     0 NA     5.74     4     2
2  1946    25 BSN   NL        8     5 0.615  2.94    24    16
3  1947    26 BSN   NL       21    10 0.677  2.33    40    35
```

The header labels Year, Age, Tm, W, L, W.L, ERA, G, GS are some variable names of the data frame; the numbers 1, 2, 3 displayed on the left give the row numbers.

The variables Age, W, L, ERA for the first 10 seasons can be displayed by use of slice() with arguments 1:10 and select() with arguments Age, W, L, ERA.

```
spahn |>
  slice(1:10) |>
  select(Age, W, L, ERA)
```

```
# A tibble: 10 x 4
     Age     W     L   ERA
   <dbl> <dbl> <dbl> <dbl>
1     21     0     0  5.74
2     25     8     5  2.94
3     26    21    10  2.33
4     27    15    12  3.71
5     28    21    14  3.07
6     29    21    17  3.16
7     30    22    14  2.98
8     31    14    19  2.98
9     32    23     7  2.1
10    33    21    12  3.14
```

Descriptive statistics of individual variables of a data frame can be obtained by use of the summarize() function in the **dplyr** package. To illustrate, we use this function to obtain some summary statistics such as the median, lower and upper quartiles, and low and high values for the ERA measure.

```
spahn |>
  summarize(
    LO = min(ERA),
```

```
    QL = quantile(ERA, .25),
    QU = quantile(ERA, .75),
    M = median(ERA),
    HI = max(ERA)
 )
```

```
# A tibble: 1 x 5
     LO    QL    QU     M    HI
  <dbl> <dbl> <dbl> <dbl> <dbl>
1   2.1  2.94  3.26  3.04  5.74
```

From this display, we see that 50% of Spahn's season ERAs fell between the lower quartile (QL) 2.94 and the upper quartile (QU) 3.26. Using the `filter()` and `select()` functions, we can find the age when Spahn had his lowest ERA by use of the following expression.

```
spahn |>
  filter(ERA == min(ERA)) |>
  select(Age)
```

```
# A tibble: 1 x 1
    Age
  <dbl>
1    32
```

Using the ERA measure, Spahn had his best pitching season at the age of 32.

2.4.3 Manipulations with data frames

The pitching variables in the `spahn` data frame are the traditional or standard pitching statistics. One can add new "sabermetric" variables to the data frame by use of the `mutate()` function in the **dplyr** package. Suppose that one wishes to measure pitching by the FIP (fielding independent pitching) statistic[1] defined by

$$FIP = \frac{13HR + 3BB - 2K}{IP}.$$

We add a new variable to a current data frame using the `mutate()` function.

```
spahn <- spahn |>
  mutate(FIP = (13 * HR + 3 * BB - 2 * SO) / IP)
```

---

[1]FIP is a measure of pitching performance dependent only on plays that do not involve fielders.

Suppose we are interested in finding the seasons where Spahn performed the best using the FIP measure. We perform this task by three functions in the **dplyr** package. The `arrange()` function sorts the data frame by the FIP measure, the `select()` function selects a group of variables, and `slice()` displays the first six rows of the data frame.

```
spahn |>
  arrange(FIP) |>
  select(Year, Age, W, L, ERA, FIP) |>
  slice(1:6)
```

```
# A tibble: 6 x 6
   Year   Age     W     L   ERA   FIP
  <dbl> <dbl> <dbl> <dbl> <dbl> <dbl>
1  1952    31    14    19  2.98 0.345
2  1953    32    23     7  2.1  0.362
3  1946    25     8     5  2.94 0.415
4  1959    38    21    15  2.96 0.675
5  1947    26    21    10  2.33 0.695
6  1956    35    20    11  2.78 0.800
```

It is interesting that Spahn's best FIP seasons occurred during the middle of his career. Also, note that Spahn had a smaller (better) FIP in 1952 compared to 1953, although his ERA was significantly larger in 1952.

Since Spahn pitched primarily for two cities, Boston and Milwaukee, suppose we are interested in comparing his pitching for the two cities. We first use the `filter()` function with a logical condition indicating that we want the `Tm` variable to be either BSN or MLN. (We introduce the logical *OR* operator |.) To compare various pitching statistics for the two teams, we use the `summarize()` function. By using the `group_by()` argument, the **spahn** data frame is grouped by `Tm`, and the mean values of the variables `W.L`, `ERA`, `WHIP`, and `FIP`) are computed for each group. The output gives the summary statistics for the Boston seasons and the Milwaukee seasons.

```
spahn |>
  filter(Tm == "BSN" | Tm == "MLN") |>
  group_by(Tm) |>
  summarize(
    mean_W.L = mean(W.L, na.rm = TRUE),
    mean_ERA = mean(ERA),
    mean_WHIP = mean(WHIP),
    mean_FIP = mean(FIP)
  )
```

```
# A tibble: 2 x 5
```

```
   Tm    mean_W.L mean_ERA mean_WHIP mean_FIP
  <chr>    <dbl>    <dbl>     <dbl>    <dbl>
1 BSN      0.577    3.36      1.33    0.792
2 MLN      0.620    3.12      1.19    0.984
```

It is interesting that Spahn's ERAs were typically higher in Boston (a mean ERA of 3.36 in Boston compared to a mean ERA of 3.12 in Milwaukee), but Spahn's FIPs were generally lower in Boston. This indicates that Spahn may have had a weaker defense or was unlucky with hits in balls in play in Boston.

2.4.4   Merging and selecting from data frames

In baseball research, it is common to have several data frames containing batting and pitching data for teams. Here we describe several ways of merging data frames and extracting a portion of a data frame that satisfies a given condition.

Suppose we read into R data frames NLbatting and ALbatting containing batting statistics for all National League and American League teams in the 2011 season. Suppose we wish to combine these data frames into a new data frame batting. To append two data frames vertically, we can use the bind_rows() function in the **dplyr** package.

```
batting <- bind_rows(NLbatting, ALbatting)
```

Suppose instead that we have read in the batting data NLbatting and the pitching data NLpitching for the NL teams in the 2011 season and we wish to match rows from one data frame to rows of the other using a particular variable as a *key*. In this case, a row of the merged data frame would contain the batting and pitching statistics for a particular team. In this case, we use the function inner_join() from the **dplyr** package where we specify the two data frames and the by argument indicates the common variable (Tm) to merge by.

```
NL <- inner_join(NLbatting, NLpitching, by = "Tm")
```

The new data frame NL contains 16 (the number of NL teams) rows and all of the variables from both the NLbatting and NLpitching data frames.

A third useful operation is choosing a subset of a data frame that satisfies a particular condition. Suppose one has the data frame NLbatting and one wishes to focus on the batting statistics for only the teams who hit over 150 home runs this season. We use the filter() function—the argument is the logical condition that describes how teams are selected.

```
NL_150 <- NLbatting |>
  filter(HR > 150)
```

The new data frame `NL_150` contains the batting statistics for the eight teams who hit over 150 home runs.

2.5 Vectors

2.5.1 Defining and computing with vectors

A fundamental structure in R is a *vector*: a sequence of values of a given type, such as numeric or character. A basic way of creating a vector is by means of the `c()` (combine) function. To illustrate, suppose we are interested in exploring the games won and lost by Spahn for the seasons after the war when he played for the Boston Braves. We create two vectors by use of the `c()` function; the games won are stored in the vector W and the games lost are stored in the vector L. The symbol `<-` is the assignment character in R. (the `=` symbol can also be used for assignment.) These lines can be directly typed into the Console window. R is case sensitive, so R will distinguish the vector L from the vector l.

```
W <- c(8, 21, 15, 21, 21, 22, 14)
L <- c(5, 10, 12, 14, 17, 14, 19)
l
```

```
Error in eval(expr, envir, enclos): object 'l' not found
```

One fundamental design principle of R is its ability to do element-by-element calculations with vectors. Suppose we wish to compute the winning percentage for Spahn for these seven seasons. We want to compute the fraction of winning games and multiply this fraction by 100 to convert it to a percentage. We create a new vector named `win_pct` by use of the basic multiplication (*) and division (/) operators:

```
win_pct <- 100 * W / (W + L)
```

We can display these winning percentages by simply typing the variable name:

```
win_pct
```

```
[1] 61.5 67.7 55.6 60.0 55.3 61.1 42.4
```

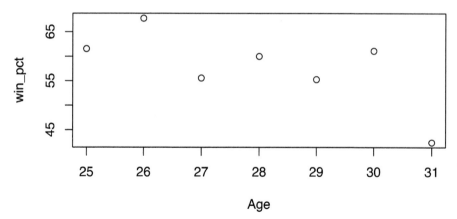

FIGURE 2.2
Scatterplot of the winning percentage against age for Warren Spahn's seasons playing for the Boston Braves.

A convenient way of creating patterned data is by use of the function `seq()`. We use this function to generate the season years from 1946 to 1952 and store the output to the variable `Year`.[2]

```
Year <- seq(from = 1946, to = 1952)
Year
```

`[1] 1946 1947 1948 1949 1950 1951 1952`

For a sequence of consecutive integer values, the colon notation will also work:

```
Year <- 1946 : 1952
```

Suppose we wish to calculate Spahn's age for these seasons. Spahn was born in April 1921 and we can compute his age by subtracting 1921 from each season value—the resulting vector is stored in the variable `Age`.

```
Age <- Year - 1921
```

We construct a simple scatterplot of Spahn's winning percentages (vertical) against his age (horizontal) by use of the `plot()` function (see Figure 2.2).

```
plot(Age, win_pct)
```

---

[2]The function `seq(a, b, s)` will generate a vector of values from a to b in steps of s.

We see that Spahn was pretty successful for most of his Boston seasons—his winning percentage exceeded 55% for six of his seven seasons.

2.5.2 Vector functions

There are many built-in R functions for vectors including `mean()` (arithmetic average), `sd()` (standard deviation), `length()` (number of vector entries), `sum()` (sum of values), `max()` (maximum value), and `sort()`. For example, one can use the `mean()` function to find the average winning percentage of Spahn during this seven-season period.

```
mean(win_pct)
```

```
[1] 57.7
```

It is actually more common to compute a pitcher's career winning percentage by dividing his cumulative win total by the total number of wins and losses. One can compute this career winning percentage by means of the following R expression.

```
100 * sum(W) / (sum(W) + sum(L))
```

```
[1] 57.3
```

One can sort the win numbers from low to high with the `sort()` function:

```
sort(W)
```

```
[1]  8 14 15 21 21 21 22
```

The `cumsum()` function is useful for displaying cumulative totals of a vector

```
cumsum(W)
```

```
[1]   8  29  44  65  86 108 122
```

We see from the output that Spahn won 8 games in the first season, 29 games in the first two seasons, and so on. The `summary()` function applied on the winning percentages displays several summary statistics of the vector values such as the extremes (low and high values), the quartiles (first and third), the median, and the mean.

```
summary(win_pct)
```

```
   Min. 1st Qu.  Median    Mean 3rd Qu.    Max.
   42.4    55.4    60.0    57.7    61.3    67.7
```

This output tells us that his median winning percentage was 60, his mean percentage was 57.66, and the entire group of winning percentages ranged from 42.42 to 67.74. Note that some of these vector functions (e.g., `sort()`, `cumsum()`) return a vector of the same length as the input vector, while others (sometimes called summary functions, e.g., `mean()`, `sd()`, `max()`) return a single value.

2.5.3 Vector index and logical variables

To extract portions of vectors, a square bracket is often used. For example, the expression

```
W[c(1, 2, 5)]
```

```
[1]  8 21 21
```

will extract the first, second, and fifth entries of the vector W. The first four values of the vector can be extracted by typing

```
W[1 : 4]
```

```
[1]  8 21 15 21
```

By use of a minus index, we remove entries from a vector. For example, if we wish to remove the first and sixth entries of W, we would type

```
W[-c(1, 6)]
```

```
[1] 21 15 21 21 14
```

A logical variable is created in R by the use of a vector together with the operations >, <, == (logical equals), and != (logical not equals). For example, suppose we are interested in the values in the winning percentage vector Win.Pct that exceed 60%.

```
win_pct > 60
```

```
[1]  TRUE  TRUE FALSE FALSE FALSE  TRUE FALSE
```

The result of this calculation is a logical vector; the output indicates that Spahn had a winning percentage exceeding 60% for the first, second, and sixth seasons (TRUE), and not exceeding 60% for the remaining seasons (FALSE). Were there any seasons where Spahn won more than 20 games and his winning percentage exceeded 60%? We use the logical & (AND) operator to find the years where W > 20 and Win.Pct > 60.

```
(W > 20) & (win_pct > 60)
```

[1] FALSE   TRUE FALSE FALSE FALSE   TRUE FALSE

The output indicates that both conditions were true for the second and sixth seasons.

By using logical variables and the square bracket notation, we can find subsets of vectors satisfying different conditions. During this period, when did Spahn have his highest winning percentage? We use

```
win_pct == max(win_pct)
```

[1] FALSE   TRUE FALSE FALSE FALSE FALSE FALSE

to create a logical vector which is true when this condition is satisfied. (Note the use of the double equal sign notion to indicate logical equality.) Then we select the corresponding year by indexing Year by this logical vector.

```
Year[win_pct == max(win_pct)]
```

[1] 1947

We see that the highest winning percentage occurred in 1947 during this period.

What seasons did the number of decisions (wins plus losses) exceed 30? We first create a logical vector based on W + L > 30, and then choose the seasons using this logical vector.

```
Year[W + L > 30]
```

[1] 1947 1949 1950 1951 1952

We see that the number of decisions exceeded 30 for the five seasons 1947, 1949, 1950, 1951, and 1952.

2.6   Objects and Containers in R

The things you create using R are called *objects*. These objects can be of different *types* such as numeric, logical, character, and integer. We have already worked with objects of types numeric and logical in the previous section. We store a number of objects into a *container*. A vector is a simple type of container where we place a number of objects of the same type, say objects that are all numeric or all logical. Here we illustrate some of the different object types and containers that we find useful in working with baseball data.

2.6.1   Character data and data frames

String variables such as the names of teams and players are stored as *characters* that are represented by letters and numbers enclosed by double quotes. As a simple example, suppose we wish to explore information about the World Series in the years 2008 through 2017. We create three character vectors NL, AL, and Winner containing abbreviations for the National League winner, the American League winner, and the league of the team that won the World Series. Note that we represent each character value by a string of letters enclosed by double quotes. We also define two numeric vectors: N_Games contains the number of games of each series, and Year gives the corresponding seasons.

```
Year <- 2008 : 2017
NL <- c("PHI", "PHI", "SFN", "SLN", "SFN",
        "SLN", "SFN", "NYN", "CHN", "LAN")
AL <- c("TBA", "NYA", "TEX", "TEX", "DET",
        "BOS", "KCA", "KCA", "CLE", "HOU")
Winner <- c("NL", "AL", "NL", "NL", "NL",
            "AL", "NL", "AL", "NL", "AL")
N_Games <- c(5, 6, 5, 7, 4, 7, 7, 5, 7, 7)
```

There are other ways to store objects besides vectors. For example, suppose we wish to display the World Series} contestants in a tabular format. A *data frame* is a rectangular grid of objects where objects within a column are the same type. More technically, a data.frame is a list (see Section 2.6.3) of vectors of the same length (but not necessarily of the same type). A data frame can be created by the data.frame() and tibble() functions, where the inputs are different vectors with associated names. Suppose we want to create a data frame containing the seasons, the National League contestants, the American League contestants, the number of games played, and the names of the World Series winners. The above vectors are used to populate the data frame, and we indicate that the names of the data frame variables are respectively Year, NL_Team, AL_Team, N_Games, and Winner. Note that the data frame is a more readable format and keeps the data organized.

```
WS_results <- tibble(
  Year = Year, NL_Team = NL, AL_Team = AL,
  N_Games = N_Games, Winner = Winner)
WS_results
```

```
# A tibble: 10 x 5
   Year NL_Team AL_Team N_Games Winner
```

```
   <int> <chr>    <chr>       <dbl> <chr>
1  2008 PHI      TBA            5 NL
2  2009 PHI      NYA            6 AL
3  2010 SFN      TEX            5 NL
4  2011 SLN      TEX            7 NL
5  2012 SFN      DET            4 NL
6  2013 SLN      BOS            7 AL
7  2014 SFN      KCA            7 NL
8  2015 NYN      KCA            5 AL
9  2016 CHN      CLE            7 NL
10 2017 LAN      HOU            7 AL
```

There are a number of R functions available for exploring character data. `str_length()`, `str_which()`, and `str_detect()` are just a few of these functions. The **stringr** packages contains many more. For example, to find the teams from New York that played in these World Series, we use `grep()` to match patterns in the text.

```
grep("NY", c(AL, NL), value = TRUE)
```

```
[1] "NYA" "NYN"
```

The `summarize()` function in the **dplyr** package together with the `group_by()` argument will summarize the data frame for each World Series league winner (variable `Winner`). To learn about the number of wins by each league in the 10 World Series, we count the rows by use of the `n()` function.

```
WS <- WS_results |>
  group_by(Winner) |>
  summarize(N = n())
WS
```

```
# A tibble: 2 x 2
  Winner     N
  <chr>  <int>
1 AL         4
2 NL         6
```

Note that the National League won 6 of these 10 World Series. One can construct a bar graph of these frequencies by use of the **ggplot2** graphics package. The `ggplot()` function indicates that we are using the data frame WS with variables `Winner` and `N`, and the `geom_col()` function says to graph each frequency value with a column (see Figure 2.3).

```
ggplot(WS, aes(x = Winner, y = N)) +
  geom_col()
```

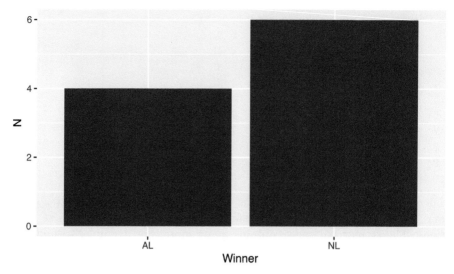

FIGURE 2.3
Bar graph of the number of wins of the American League and National League teams in the World Series between 2003 and 2012.

Equivalently, we could let **ggplot2** do the summarizing for us by using the geom_bar() function.

```
ggplot(WS_results, aes(x = Winner)) +
  geom_bar()
```

2.6.2 Factors

A *factor* is a special way of representing character data. To motivate the consideration of factors, suppose we construct a frequency table of the National League representatives to the World Series in the character vector NL_Team.

```
WS_results |>
  group_by(NL_Team) |>
  summarize(N = n())
```

```
# A tibble: 6 x 2
  NL_Team     N
  <chr>    <int>
1 CHN          1
2 LAN          1
3 NYN          1
4 PHI          2
```

```
5 SFN           3
6 SLN           2
```

Note that R will organize the teams alphabetically (from CHN to STL) in the frequency table. It may be preferable to organize the teams by the division (East, Central, and West). We can change the organization of the team labels by converting this character type to a factor.

We redefine the NL_Team variable by means of the mutate() function (in the dplyr package) and the factor() function. The basic arguments to factor() are the vector of data to be converted and a vector levels that gives the ordered values of the variable. Here we list the values ordered by the East, Central, and West divisions.

```
WS_results <- WS_results |>
  mutate(
    NL_Team = factor(
      NL_Team,
      levels = c("NYN", "PHI", "CHN", "SLN", "LAN", "SFN")
    )
  )
```

One can understand how factor variables are stored by using the str() function to examine the structure of the variable NL_Team.

```
str(WS_results$NL_Team)
```

```
Factor w/ 6 levels "NYN","PHI","CHN",..: 2 2 6 4 6 4 6 1 3 5
```

We see that a factor variable is actually encoded by integers (2, 2, 6, ...) where the levels are the team names. If we reconstruct the table by use of the summarize() function, grouping by the variable NL_Team, we obtain the same frequencies as before, but the teams are now listed in the order specified in the factor() function.

```
WS_results |>
  group_by(NL_Team) |>
  summarize(N = n())
```

```
# A tibble: 6 x 2
  NL_Team      N
  <fct>    <int>
1 NYN          1
2 PHI          2
3 CHN          1
4 SLN          2
```

```
5 LAN          1
6 SFN          3
```

Many R functions require the use of factors, and the use of factors gives one finer control on how character labels are displayed in output and graphs.

2.6.3 Lists

A container such as a vector requires that data values have the same type. For example, vectors contain all numeric data or all character data; one cannot mix numeric and character data in a single vector. A data frame is an example of a container that contain vectors of different types and a *list* is a general way of storing "mixed" data. As noted previously, a data.frame is a special case of a list in which every element is a vector, and all of those vectors have the same length. In general, the elements of a list can be any R object. To illustrate, suppose we wish to collect the league that won the World Series (a character type), the number of games played (a numeric type), and a short description (a character type) into a single variable. Using the list() function, we create a new list world_series with components Winner, Number.Games, and Seasons.

```
world_series <- list(
  Winner = Winner,
  Number_Games = N_Games,
  Seasons = "2008 to 2017"
)
```

Once a list such as world_series is defined, there are different ways of accessing the different components. If we wish to display the number of games played Number.Games, we can use the list variable name together with the $ symbol and the component name.

```
world_series$Number_Games
```

```
[1] 5 6 5 7 4 7 7 5 7 7
```

Or we can use the double square brackets to display the second component of the list.

```
world_series[[2]]
```

```
[1] 5 6 5 7 4 7 7 5 7 7
```

The pluck() function from the **purrr** package also extracts elements from a list.

```
pluck(world_series, "Number_Games")
```

```
[1] 5 6 5 7 4 7 7 5 7 7
```

As an alternative, we can use the single square brackets with the name of the component in quotes.

```
world_series["Number_Games"]
```

```
$Number_Games
 [1] 5 6 5 7 4 7 7 5 7 7
```

Note that the first three options return vectors and the fourth option returns a list with the single component Number_Games.

Since a data.frame is a list, the dollar sign operator can be used to extract a vector from a data.frame as well. The pull() function from the **dplyr** package achieves the same effect.

```
WS_results$NL_Team
```

```
 [1] PHI PHI SFN SLN SFN SLN SFN NYN CHN LAN
Levels: NYN PHI CHN SLN LAN SFN
```

```
pull(WS_results, NL_Team)
```

```
 [1] PHI PHI SFN SLN SFN SLN SFN NYN CHN LAN
Levels: NYN PHI CHN SLN LAN SFN
```

Many R functions return lists of data of different types, so it is important to know how to access the components of a list. Also we will see that lists provide a convenient way of collecting information of different types (character, numeric, logical, factors) about teams and players.

2.7 Collection of R Commands

2.7.1 R scripts

The R expressions described in the previous sections can be typed directly in the Console window and any output will be directly displayed in that window. Alternatively, R expressions can be stored in a text file called an R script and executed as a group.

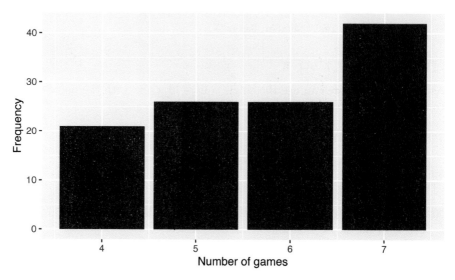

FIGURE 2.4
Bar graph of the number of games played in best of seven World Series since 1903.

Suppose we wish to run the following R commands. The data frame `SeriesPost` in the **Lahman** package contains information about all MLB playoff games—two of the variables are `wins` and `losses`, the number of games won and lost by the winning team in the series. First, we create a new data frame `ws` containing data from all of the World Series with fewer than 8 games played. Using the **ggplot2** package, we construct the bar graph of the number of games played in all "best of seven" World Series shown in Figure 2.4.

```
library(Lahman)
ws <- SeriesPost |>
  filter(yearID >= 1903, round == "WS", wins + losses < 8)
ggplot(ws, aes(x = wins + losses)) +
  geom_bar(fill = crcblue) +
  labs(x = "Number of games", y = "Frequency")
```

A convenient way to run R scripts is through the text window in the upper-left window of the RStudio environment. The R commands above are typed in this window and the script is executed by selecting these lines and pressing Control-Enter (in a Linux or Windows operating system) or Command-Enter (in a Macintosh operating system). The screenshot in Figure 2.5 shows the result of executing this R script. The R output is displayed in the lower-left Command window. In the Workspace window (upper-right), we see that the data frame `ws` has been created. In the Plots window (lower-right), we see the bar graph as a result of the graphics functions.

Another way of running an R script is by saving the commands in a file, and then using the `source()` function to load this file into R. Suppose that a file

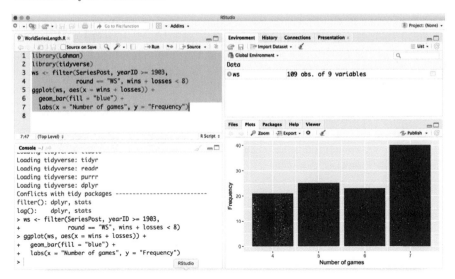

FIGURE 2.5
Snapshot of the RStudio interface after executing commands from an R script.

with the above commands has been saved in the file `WorldSeriesLength.R` in the `scripts` subdirectory of the current working directory. (See Section 2.8.1 for information about changing the working directory.) Then one can execute this file by typing in the Console window:

```
source(here::here("scripts/WorldSeriesLength.R"), echo = TRUE)
```

The `echo = TRUE` argument is used so that the R output is displayed in the Console window.

2.7.2 R functions

We have illustrated the use of a number of R built-in packages. One attractive feature of R is the capability to create one's own functions to implement specific computations and graphs of interest.

As a simple example, suppose you are interested in writing a function to compute a player's home run rates for a collection of seasons. One inputs a vector `age` of player ages, a vector `hr` of home run counts, and a vector `ab` of at-bats. You want the function to compute the player's home run rates (as a percentage, rounded to the nearest tenth), and output the ages and rates in a form amenable to graphing.

The following function `hr_rates()` will perform the desired calculations. All functions start with the syntax `name_of_function <- function(arguments)`, where arguments is a list of input variables.

All of the work in the function goes inside the curly brackets that follow. The result of the last line of the function is returned as the output. In our example, the name of the function is `hr_rates` and there are three vector inputs `age`, `hr`, and `ab`. The `round()` function is used to compute the home run rates.[3] The output of this function is a list with two components: `x` is the vector of ages, and `y` is the vector of home run rates.

```
hr_rates <- function(age, hr, ab) {
  rates <- round(100 * hr / ab, 1)
  list(x = age, y = rates)
}
```

To use this function, first it needs to be read into R. This can be done by entering it directly into the Console window, or by saving the function in a file, say `hr_rates.R`, and reading it into R by the `source()` function. (This function is also available in the **abdwr3edata** package.)

```
source(here::here("scripts/hr_rates.R"))
```

We illustrate using this function on some home run data for Mickey Mantle for the seasons 1951 to 1961. We enter Mantle's home run counts in the vector `HR`, the corresponding at-bats in `AB`, and the ages in `Age`. We apply the function `hr_rates()` with inputs `Age`, `HR`, `AB`, and the output is a list with Mantle's ages and corresponding home run rates.

```
HR <- c(13, 23, 21, 27, 37, 52, 34, 42, 31, 40, 54)
AB <- c(341, 549, 461, 543, 517, 533, 474, 519, 541, 527, 514)
Age <- 19 : 29
hr_rates(Age, HR, AB)
```

```
$x
 [1] 19 20 21 22 23 24 25 26 27 28 29
```

```
$y
 [1]  3.8  4.2  4.6  5.0  7.2  9.8  7.2  8.1  5.7  7.6 10.5
```

One can easily construct a scatterplot (not shown here) of Mantle's rates against age by the `plot()` function on the output of the function.

```
plot(hr_rates(Age, HR, AB))
```

Verify that Mantle's home run rates rose steadily in the first six seasons of his career.

---

[3]The expression `round(x, n)` rounds `x` to `n` decimal places.

2.8 Reading and Writing Data in R

2.8.1 Importing data from a file

Generally it is tedious to input data manually into R. For the large data files that we will be working with in this book, it will be necessary to import these files directly into R. We illustrate this importing process using the complete pitching profile of Spahn.

We created the file **spahn.csv** containing Spahn's pitching statistics and placed the file in the current working directory. One can check the location of the current working directory in R by means of typing `getwd()` in the Console window:

```
getwd()
```

```
[1] "/home/bbaumer/Dropbox/git/abdwr3e/book"
```

In RStudio, one can change the working directory by selecting the "Change Working Directory" option on the Tools menu or by use of the `setwd()` function. One can easily import this dataset in RStudio by pressing the "Import Dataset" button in the top right window. You select the "From Text File" option and find the dataset of interest. After you select the file, Figure 2.6 shows a snapshot of the Import Dataset window. One sees the input file and also the format of the

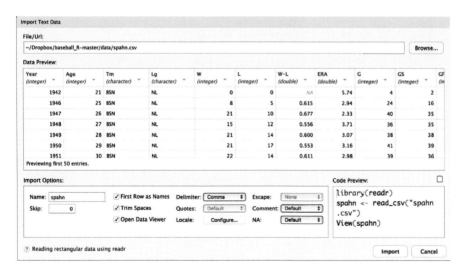

FIGURE 2.6
Snapshot of the Import Dataset window in the RStudio interface.

data that will be saved into R. It is important to check the button that the file contains a heading, which means the first line of the input file contains the variable names.

An alternative method of importing data from a file uses the `read_csv()` function from the **readr** package. This function assumes the file is stored in a "comma separated value" format, where different values on a single row are separated by commas. For our example, the following R expression reads the comma separated value file `spahn.csv` stored in the `data` directory in the current working directory and saves the data into a data frame with name `spahn`.

```
spahn <- read_csv(here::here("data/spahn.csv"))
```

2.8.2 Saving datasets

We have seen that it is straightforward to read comma-delimited data files (csv format) into R by use of the `read_csv()` function. Similarly, we can use the `write_csv()` function from the **readr** package to save datasets in R in the CSV format.

We return to the Mickey Mantle example where we have vectors of home run counts, at-bats, and ages, and we use the user-defined function `hr_rates()` to compute home run rates. We create a data frame `Mantle` combining the vectors `Age`, `HR`, `AB`, and the y component of the list `hr_rates` using the `tibble()` function.

```
mantle_hr_rates <- hr_rates(Age, HR, AB)
Mantle <- tibble(
  Age, HR, AB, Rates = mantle_hr_rates$y
)
```

We use the `write_csv()` function to save the data to the current working directory. This function has two arguments: the R object `Mantle` that we wish to save, and the output file path `data/mantle.csv`.

```
write_csv(Mantle, here::here("data/mantle.csv"))
```

It is good to confirm (using `list.files()`) that a new file `mantle.csv` exists in the current working directory.

```
list.files(here::here("data"), pattern = "mantle")
```

```
[1] "mantle.csv"
```

2.9   Packages

Many useful functions are available through the base R system. However, one attractive feature of R is the availability of collections of functions and datasets in R packages. Currently, there are over 20,000 packages contributed by R users available on the R website (https://cran.r-project.org/), and these packages expand the capabilities of the R system. In this book, we focus on a few contributed packages that we find useful in our baseball work.

To illustrate installing and loading an R package, the **Lahman** package contains the data files from the Lahman database described in Section 1.2. Assuming one is connected to the Internet, one can install the current version of this package into R by means of the command

```
install.packages("Lahman")
```

Alternately, one can install packages by use of the Install Packages button on the Package tab in RStudio.

After a package has been installed, then one needs to load the package into R to have access to the functions and datasets. For example, to load the new package **Lahman**, one types

```
library(Lahman)
```

To confirm that the package has been loaded correctly, we use the `help()` function to learn about the dataset `Batting` in the **Lahman** package. (A general discussion of the `help()` function is given in Section 2.11.)

```
?Batting
```

When one launches R, one needs to load the packages that are not automatically loaded in the system.

2.10   Splitting, Applying, and Combining Data

In many situations, one is interested in splitting a data frame into parts, applying some operation on each part, and then combining the results in a new data frame. This type of "split, apply, combine" operation is facilitated using the `group_by()` and `summarize()` functions in the **dplyr** package. Here we illustrate this process on the **Lahman** batting database. In this work, we review some other handy data frame manipulation functions previously discussed.

Suppose we are interested in looking at the great home run hitters in baseball history. Specifically, we want to answer the question "Who hit the most home runs in the 1960s?"

We begin by loading in the **Lahman** package.

```
library(Lahman)
```

Remember that the data frame `Batting` contains the season batting statistics for all players in baseball history. Since we are focusing on the 1960s, the `filter()` function is used to select batting data only for the seasons between 1960 and 1969, creating the new data frame `Batting_60`.

```
Batting_60 <- Batting |>
  filter(yearID >= 1960, yearID <= 1969)
```

Suppose we would like to compute the total number of home runs for each player in the data frame `Batting_60`. The key variables are the player identification code `playerID` and the home run count `HR`. We want to split the data frame by each player id, and then compute the sum of home runs for each player. In the code below, the splitting is accomplished by the `group_by()` argument, and the sum of home runs is computed using the `summarize()` function.

```
hr_60 <- Batting_60 |>
  group_by(playerID) |>
  summarize(HR = sum(HR))
```

The output is a data frame `hr_60` containing two variables, `playerID` and the home run count `HR`.

Using the `arrange()` function with the `desc()` argument, we sort this data frame in descending order so that the best home run hitters are on the top, and display the first four lines of this data frame.

```
hr_60 |>
  arrange(desc(HR)) |>
  slice(1:4)
```

```
# A tibble: 4 x 2
  playerID      HR
  <chr>      <int>
1 killeha01    393
2 aaronha01    375
```

```
3 mayswi01     350
4 robinfr02    316
```

The most prolific home run hitters in the 1960s were Harmon Killebrew, Hank Aaron, Willie Mays, and Frank Robinson.

We could also perform this sequence of operations in a single pipeline. This has the advantage of not cluttering the workspace with unnecessary intermediate data sets.

```
Batting |>
  filter(yearID >= 1960, yearID <= 1969) |>
  group_by(playerID) |>
  summarize(HR = sum(HR)) |>
  arrange(desc(HR)) |>
  slice(1:4)
```

2.10.1 Iterating using map()

A key competency in data science is the ability to iterate an analytic operation over sequences of inputs. Pursuant to the discussion above, suppose now that we want to identify the player who hit the most home runs in each decade across baseball history.

First, we write a simple function that will take a data frame of batting statistics and return one row corresponding to the player with the most home runs in that set. This requires only a slight modification of the previous code.

```
hr_leader <- function(data) {
  data |>
    group_by(playerID) |>
    summarize(HR = sum(HR)) |>
    arrange(desc(HR)) |>
    slice(1)
}
```

Next, we need to split the `Batting` data frame into pieces based on the decade. This information is not stored in `Batting`, so we use `mutate()` to create a new variable called `decade` that computes the first year of the decade for every value of `yearID`. Then, we use the `group_by()` function to organize `Batting` into pieces that share common values of `decade`. This results in a grouped data frame, with each group corresponding to a single decade.

```
Batting_decade <- Batting |>
  mutate(decade = 10 * floor(yearID / 10)) |>
  group_by(decade)
```

Next, we use the `group_keys()` function to retrieve a vector of the first year in each decade. We'll need these later.

```
decades <- Batting_decade |>
  group_keys() |>
  pull("decade")
decades
```

```
 [1] 1870 1880 1890 1900 1910 1920 1930 1940 1950 1960 1970 1980
[13] 1990 2000 2010 2020
```

Finally, we use the `group_split()` function to break `Batting_decade` into pieces, and the `map()` function from the **purrr** package to apply our `hr_leader()` function to each of those data frames. The `set_names()` function and the `.id` argument to `bind_rows()` ensure that the variable displaying the first year of the decade gets the right name.

```
Batting_decade |>
  group_split() |>
  map(hr_leader) |>
  set_names(decades) |>
  bind_rows(.id = "decade")
```

```
# A tibble: 16 x 3
   decade playerID       HR
   <chr>  <chr>       <int>
 1 1870   pikeli01       21
 2 1880   stoveha01      89
 3 1890   duffyhu01      83
 4 1900   davisha01      67
 5 1910   cravaga01     116
 6 1920   ruthba01      467
 7 1930   foxxji01      415
 8 1940   willite01     234
 9 1950   snidedu01     326
10 1960   killeha01     393
11 1970   stargwi01     296
12 1980   schmimi01     313
13 1990   mcgwima01     405
14 2000   rodrial01     435
15 2010   cruzne02      346
16 2020   judgeaa01     110
```

Note that this confirms our previous finding that Harmon Killebrew hit the most home runs in the 1960s, but also informs us that Babe Ruth holds the record for most home runs hit in a single decade (467 in the 1920s), followed by Alex Rodriguez (who hit 435 home runs in the 2000s).

2.10.2 Another example

Using the same `Batting` data frame of season batting statistics, suppose we are interested in collecting the career at-bats, career home runs, and career strikeouts for all players in baseball history with at least 5000 career at-bats. Both home runs and strikeouts are of interest since we suspect there may be some association between a player's strikeout rate (defined by SO/AB) and his home run rate HR/AB.

This operation is done in two steps. First, we create a new data frame consisting of the career `AB`, `HR`, and `SO` for all batters. Second, by use of the `filter()` function, the batting seasons are selected from the data frame for the players with 5000 AB.

The function `summarize()` in the **dplyr** package is useful for the first operation. We want to compute the sum of `AB` over the seasons of a player's career. The `group_by()` function indicates we wish to split the `Batting` data frame by the `playerID` variable, and `tAB = sum(AB, na.rm = TRUE)` indicates we wish to summarize each data frame "part" by computing the sum of the `AB`. (Some of the `AB` values will be missing and coded as `NA`, and the `na.rm = TRUE` will remove these missing values before taking the sum.) The new data frame `long_careers` contains the career AB for all players.

```
long_careers <- Batting |>
  group_by(playerID) |>
  summarize(
    tAB = sum(AB, na.rm = TRUE),
    tHR = sum(HR, na.rm = TRUE),
    tSO = sum(SO, na.rm = TRUE)
  )
```

Now that we have this new variable `tAB`, one can now use the `filter()` function to choose only the season batting statistics for the players with 5000 AB.

```
Batting_5000 <- long_careers |>
  filter(tAB >= 5000)
```

The resulting data frame `Batting_5000` contains the career AB, HR, and SO for all batters with at least 5000 career AB. To confirm, the first six lines of the data frame are displayed by the `slice()` function.

```
Batting_5000 |>
  slice(1:6)
```

```
# A tibble: 6 x 4
  playerID     tAB   tHR    tSO
  <chr>      <int> <int>  <int>
1 aaronha01  12364   755   1383
2 abreubo01   8480   288   1840
3 adamssp01   5557     9    223
4 adcocjo01   6606   336   1059
5 alfoned01   5385   146    617
6 allendi01   6332   351   1556
```

Is there an association between a player's home run rate and his strikeout rate? Using the `geom_point()` function, we construct a scatterplot of `HR/AB` and `SO/AB`. Using the `geom_smooth()` function, we add a smoothing curve (see Figure 2.7).

```
ggplot(Batting_5000, aes(x = tHR / tAB, y = tSO / tAB)) +
    geom_point() + geom_smooth(color = crcblue)
```

It is clear from the graph that batters with higher home run rates tend to have higher strikeout rates.

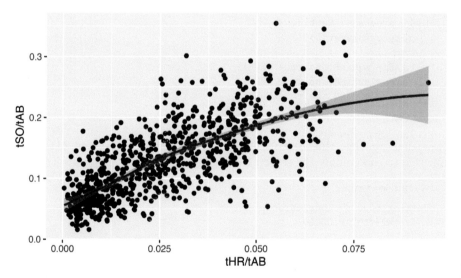

FIGURE 2.7
Scatterplot of the homerun rates and strikeout rates of all players with at least 5000 career at-bats. A smoothing curve is added to the plot to show that home run rates and strikeout rates have a positive association.

2.11 Getting Help

The Help menu in RStudio provides general documentation about the R system (see the R Help option). From the Help menu, we also find general information about the RStudio system such as keyboard shortcuts. In addition, R contains an online help system providing documentation on R functions and datasets. For example, suppose you wish to learn about the `geom_point()` function that constructs a scatterplot, a statistical graphical display discussed in Chapter 3. By typing in the Console window a question mark followed by the function name,

```
?geom_point
```

you see a long description of this function including all of the possible function arguments. To find out about related functions, one can preface `geom_point` by two question marks to find all objects that contain this character string:

```
??geom_point
```

RStudio provides an additional online help system that is especially helpful when one does not know the exact spelling of an R function. For example, suppose I want to construct a dot chart, but all I know is that the function contains the string "geom". In the Console window, I type "geom" followed with a Tab. RStudio will complete the code, forming `geom_point()` and showing an abbreviated description of the function. In the case where the character string does not uniquely define the function, RStudio will display all of the functions with that string.

2.12 Further Reading

R is an increasingly popular system for performing data analysis and graphics, and a large number of books are available which introduce the system. The manual "An Introduction to R" (Venables, Smith, and R Development Core Team 2011), available through the R and RStudio systems, provides a broad overview of the R language, and the manual "R Data Import/Export" provides an extended description of R capabilities to import and export datasets. Kabacoff (2010) and the accompanying website https://www.statmethods.net provide helpful advice on specific R functions on data input, data management, and graphics. Albert and Rizzo (2012) provide an example-based introduction to R, where different chapters are devoted to specific statistics topics such as exploratory fitting, modeling, graphics, and simulation.

Several introductory data science textbooks use R extensively. Wickham, Çetinkaya-Rundel, and Grolemund (2023) is a free, online book that illustrates how the tidyverse tools can be used for data science. Benjamin S. Baumer, Kaplan, and Horton (2021b) is a data science textbook that employs tidyverse-compliant code. Ismay and Kim (2019) is another free, online textbook that uses the tidyverse to explicate statistical and data science concepts.

2.13   Exercises

1. Top Base Stealers in the Hall of Fame

The following table gives the number of stolen bases (SB), the number of times caught stealing (CS), and the number of games played (G) for nine players currently inducted in the Hall of Fame.

Player	SB	CS	G
Rickey Henderson	1406	335	3081
Lou Brock	938	307	2616
Ty Cobb	896	178	3035
Tim Raines	808	146	2502
Eddie Collins	741	173	2826
Max Carey	738	92	2476
Joe Morgan	689	162	2649
Ozzie Smith	580	148	2573
Barry Bonds	514	141	2986
Ichiro Suzuki	509	117	2653
Luis Aparicio	506	136	2601
Paul Molitor	504	131	2683
Roberto Alomar	474	114	2379

a. In R, place the stolen base, caught stealing, and game counts in the vectors SB, CS, and G.

b. For all players, compute the number of stolen base attempts SB + CS and store in the vector SB.Attempt.

c. For all players, compute the success rate Success.Rate = SB / SB.Attempt.

d. Compute the number of stolen bases per game SB.Game = SB / Game.

e. Construct a scatterplot of the stolen bases per game against the success rates. Are there particular players with unusually high or low stolen base success rates? Which player had the greatest number of stolen bases per game?

2. Character, Factor, and Logical Variables in R

Suppose one records the outcomes of a batter in ten plate appearances:

Single, Out, Out, Single, Out, Double, Out, Walk, Out, Single

 a. Use the `c()` function to collect these outcomes in a character vector `outcomes`.
 b. Use the `table()` function to construct a frequency table of `outcomes`.
 c. In tabulating these results, suppose one prefers the results to be ordered from least-successful to most-successful. Use the following code to convert the character vector `outcomes` to a factor variable `f.outcomes`.

```
f.outcomes <- factor(
  outcomes,
  levels = c("Out", "Walk", "Single", "Double")
)
```

Use the `table()` function to tabulate the values in `f.outcomes`. How does the output differ from what you saw in part (b)?

 d. Suppose you want to focus only on the walks in the plate appearances. Describe what is done in each of the following statements.

```
outcomes == "Walk"
sum(outcomes == "Walk")
```

3. Pitchers in the 350-Wins Club

The following table lists all nine pitchers who have won at least 350 career wins.

Player	W	L	SO	BB
Pete Alexander	373	208	2198	951
Roger Clemens	354	184	4672	1580
Pud Galvin	365	310	1807	745
Walter Johnson	417	279	3509	1363
Greg Maddux	355	227	3371	999
Christy Mathewson	373	188	2507	848
Kid Nichols	362	208	1881	1272
Warren Spahn	363	245	2583	1434
Cy Young	511	315	2803	1217

 a. In R, place the wins and losses in the vectors W and L, respectively. Also, create a character vector `Name` containing the last names of these pitchers.

b. Compute the winning percentage for all pitchers defined by $100 \times W/(W+L)$ and put these winning percentages in the vector `win_pCT`.
c. By use of the command

```
wins_350 <- tibble(Name, W, L, win_pCT)
```

create a data frame `wins_350` containing the names, wins, losses, and winning percentages. d. By use of the `arrange()` function, sort the data frame `wins_350` by winning percentage. Among these pitchers, who had the largest and smallest winning percentages?

4. Pitchers in the 350-Wins Club, Continued

a. In R, place the strikeout and walk totals from the 350 win pitchers in the vectors SO and BB, respectively. Also, create a character vector `Name` containing the last names of these pitchers.
b. Compute the strikeout-walk ratio by SO/BB and put these ratios in the vector `SO.BB.Ratio`.
c. By use of the command

```
SO.BB <- tibble(Name, SO, BB, SO.BB.Ratio)
```

create a data frame `SO.BB` containing the names, strikeouts, walks, and strikeout-walk ratios. d. By use of the `filter()` function, find the pitchers who had a strikeout-walk ratio exceeding 2.8. e. By use of the `arrange()` function, sort the data frame by the number of walks. Did the pitcher with the largest number of walks have a high or low strikeout-walk ratio?

5. Pitcher Strikeout/Walk Ratios

a. Read the **Lahman** `Pitching` data into R.
b. The following script computes the cumulative strikeouts, cumulative walks, mid career year, and the total innings pitched (measured in terms of outs) for all pitchers in the data file.

```
career_pitching <- Pitching |>
  group_by(playerID) |>
  summarize(
    SO = sum(SO, na.rm = TRUE),
    BB = sum(BB, na.rm = TRUE),
    IPouts = sum(IPouts, na.rm = TRUE),
    midYear = median(yearID, na.rm = TRUE)
  )
```

This new data frame is named `career_pitching`. Run this code and use the `inner_join()` function to merge the `Pitching` and `career_pitching` data frames.

c. Use the `filter()` function to construct a new data frame `career_10000` consisting of data for only those pitchers with at least 10,000 career IPouts.

d. For the pitchers with at least 10,000 career IPouts, construct a scatterplot of mid career year and ratio of strikeouts to walks. Comment on the general pattern in this scatterplot.

3

Graphics

3.1 Introduction

To illustrate methods for creating graphs in R in the **ggplot2** package (Wickham 2016b), consider all the career batting statistics for the current members of the Hall of Fame. The data frame `hof_batting` in the **abdwr3edata** package contains the career batting statistics for this group. We copy these data into a data frame named `hof`.

```
library(tidyverse)
library(abdwr3edata)
hof <- hof_batting
```

If we remove the pitchers' batting statistics from the dataset, one has statistics for 167 non-pitchers. The type of graph we use depends on the measurement scale of the variable. There are two fundamental data types—measurement and categorical—which are represented in R as numeric and character variables. We initially describe graphs for a single character variable and a single numeric variable, and then describe graphical displays helpful for understanding relationships between the variables. Using the **ggplot2** system, it is easy to modify the attributes of a graph by adding labels and changing the style of plotting symbols and lines. After describing the graphical methods, we describe the process of creating graphs for two home run stories. In Section 3.7, we compare the home run career progress of four great sluggers in baseball history, while Section 3.8 we illustrate the famous home run race of Mark McGwire and Sammy Sosa during the 1998 season.

3.2 Character Variable

3.2.1 A bar graph

The Hall-of-Famers played during different eras of baseball; one common classification of eras is "19th Century" (up to the 1900 season), "Dead Ball" (1901 through 1919), "Lively Ball" (1920 though 1941), "Integration" (1942 through 1960), "Expansion" (1961 through 1976), "Free Agency" (1977 through 1993), and "Long Ball" (after 1993). We want to create a new character variable Era giving the era for each player. First, we define a player's mid career (variable MidCareer) as the average of his first and last seasons in baseball. We then use the mutate() and cut() functions to create the new factor variable Era—the arguments to the function are the numeric variable to be discretized, the vector of cut points, and the vector of labels for the categories of the factor variable.

```
hof <- hof |>
  mutate(
    MidCareer = (From + To) / 2,
    Era = cut(
      MidCareer,
      breaks = c(1800, 1900, 1919, 1941, 1960, 1976, 1993, 2050),
      labels = c(
        "19th Century", "Dead Ball", "Lively Ball", "Integration",
        "Expansion", "Free Agency", "Long Ball"
      )
    )
  )
```

A frequency table of the variable Era can be constructed using the summarize() function with the n() function. Below, we store that output in the data frame hof_eras.

```
hof_eras <- hof |>
  group_by(Era) |>
  summarize(N = n())
hof_eras
```

```
# A tibble: 7 x 2
  Era                N
  <fct>          <int>
1 19th Century      18
2 Dead Ball         19
3 Lively Ball       46
4 Integration       24
5 Expansion         23
6 Free Agency       22
7 Long Ball         15
```

We construct a bar graph from those data using the geom_bar() function in **ggplot2**.

The aes() function defines aesthetics. There are mappings between visual elements on the plot and variables in the data frame. Here we map the character vector Era to the x aesthetic, which defines horizontal positioning. Figure 3.1

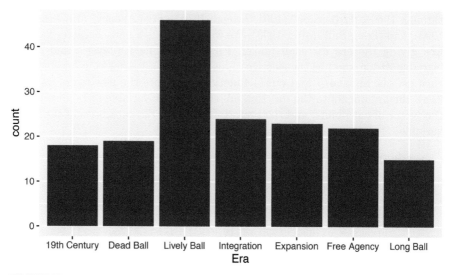

FIGURE 3.1
Bar graph of the era of the Hall of Fame non-pitchers.

shows the resulting graph. We see that a large number of these Hall of Fame players played during the Lively Ball era.

```
ggplot(hof, aes(x = Era)) + geom_bar()
```

3.2.2   Add axes labels and a title

As good practice, graphs should have descriptive axes labels and a title for describing the main message of the display. In the **ggplot2** package, the functions `xlab()` and `ylab()` add horizontal and vertical axis labels and the `ggtitle()` function adds a title. In the following code to construct a bar graph, we add the labels "Baseball Era" and "Frequency" and add the title "Era of the Nonpitching Hall of Famers". The enhanced plot is shown in Figure 3.2.

```
ggplot(hof, aes(Era)) +
  geom_bar() +
  xlab("Baseball Era") +
  ylab("Frequency") +
  ggtitle("Era of the Nonpitching Hall of Famers")
```

3.2.3   Other graphs of a character variable

There are alternative graphical displays for a table of frequencies of a character variable. For the data frame of era frequencies, we use the function `geom_point()` to construct a Cleveland-style (Cleveland 1985) dot plot shown

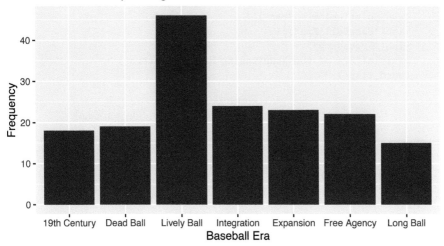

FIGURE 3.2
Era of the non-pitching Hall of Famers.

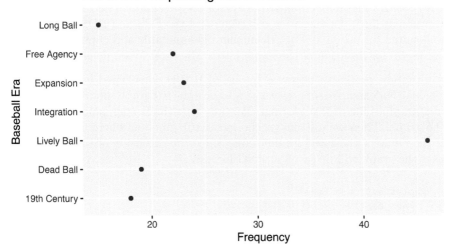

FIGURE 3.3
Dot plot of era of the Hall of Fame non-pitchers.

in Figure 3.3. A dot plot is helpful when there are a large number of categories of the character vector. The dots are colored red by the `color = "red"` argument in `geom_plot()`.

```
ggplot(hof_eras, aes(Era, N)) +
  geom_point(color = "red") +
  xlab("Baseball Era") +
  ylab("Frequency") +
  ggtitle("Era of the Nonpitching Hall of Famers") +
  coord_flip()
```

3.3 Saving Graphs

After a graph is produced in R, it is straightforward to export it to one of the usual graphics formats so that it can be used in a document, blog, or website. We outline the steps for saving graphs in the RStudio interface.

If a graph appears in the Plots window of RStudio, then the Export menu allows one to "Save Plot as Image", "Save Plot as PDF", or "Copy Plot to the Clipboard". If one chooses the "Save Plot as Image" option, then by choosing an option from a drop-down menu, one can save the graph in PNG, JPEG, TIFF, BMP, metafile, clipboard, SVG, or EPS formats. The PNG format is convenient for uploading to a web page, and the EPS and PDF formats are well-suited for use in a LATEX document. The metafile and clipboard options are useful for insertion of the graph into a Microsoft Word document.

Alternately, plots can be saved by use of R functions typed in the Console window. For example, suppose we wish to save the bar graph shown in Figure 3.2 in a graphics file of PNG format. We first type the R commands to produce the graph. Then we use the special `ggsave()` function where the argument is the name of the saved graphics file. Since the extension of the filename is **png**, the graph will be saved in PNG format.

```
ggplot(hof, aes(Era)) +
  geom_bar() +
  xlab("Baseball Era") +
  ylab("Frequency") +
  ggtitle("Era of the Nonpitching Hall of Famers")
ggsave("bargraph.png")
```

If we look at the current directory, we will see a new file `bargraph.png` containing the image in PNG format. The graph can be saved in alternative graphics formats by use of different extensions. For example the argument to `ggsave()` would be **pdf** if we wished to save the graph in PDF format or **jpeg** if we wanted save in the JPEG format.

Other methods of saving graphs are useful if one wishes to save a number of graphs in a single file. For example, one can use the **patchwork** library to combine more than one ggplot into a single ggplot object. This composite plot can then be saved using the aforementioned ggsave() command. For example, if one types:

```
library(patchwork)
p1 <- ggplot(hof, aes(Era)) + geom_bar()
p2 <- ggplot(hof_eras, aes(Era, N)) + geom_point()
p1 + p2
ggsave("graphs.pdf")
```

then the bar graph and the dot plot graph will be saved together in the PDF file graphs.pdf.

3.4 Numeric Variable: One-Dimensional Scatterplot and Histogram

When one collects a numeric variable such as a batting average, an on-base percentage, or an OPS from a group of players, one typically wants to learn about its distribution. For example, if we examine OPS values for the nonpitcher Hall of Fame inductees, we are interested in learning about the general shape of the OPS values. For example, is the distribution of OPS values symmetric, or is it right or left skewed? Also we are interested in learning about the typical or representative Hall of Fame OPS value, and how the OPS values are spread out. Graphical displays provide a quick visual way of studying distributions of baseball statistics.

For a single numeric variable, two useful displays for visualizing a distribution are the one-dimensional scatterplot and the histogram. A one-dimensional scatterplot is basically a number line graph, where the values of the statistics are plotted over a number line ranging over all possible values of the variable. We construct a graph of the OPS values for the Hall of Fame inductees in **ggplot2** by the geom_jitter() function. In the data frame hof, the OPS is mapped to the x aesthetic and the dummy variable y is set to a constant value. The theme elements are chosen to remove the tick marks, text, and title from the y-axis.

```
ggplot(hof, aes(x = OPS, y = 1)) +
  geom_jitter(height = 0.2) +
  ylim(0, 2) +
  theme(
```

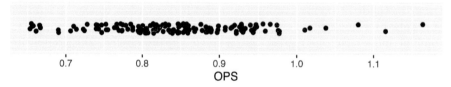

FIGURE 3.4
One-dimensional scatterplot of the OPS values of the Hall of Fame players.

```
      axis.title.y = element_blank(),
      axis.text.y = element_blank(),
      axis.ticks.y = element_blank()
  ) +
  coord_fixed(ratio = 0.03)
```

The resulting graph is shown in Figure 3.4. One sees that most of the OPS values fall between 0.700 and 1.000, but there are a few unusually high values that could merit further exploration.

A second graphical display for a numeric variable is a histogram where the values are grouped into bins of equal width and the bin frequencies are displayed as non-overlapping bars over the bins. A histogram of the OPS values is constructed in the **ggplot2** system by use of the function `geom_histogram()`. The only aesthetic mapping is to the variable `OPS` (see Figure 3.5).

```
ggplot(hof, aes(x = OPS)) +
  geom_histogram()
```

One issue in constructing a histogram is the choice of bins, and the function `geom_histogram()` will typically make reasonable choices for the bins to produce a good display of the data distribution. One can select one's own bins in `geom_histogram()` by use of the argument `breaks`. For example, if one wanted to choose the alternative bin endpoints $0.4, 0.5, \ldots, 1.2$, then one could construct the histogram by the following code (see Figure 3.6). By use of the `color` and `fill` arguments, the lines of the bars are colored white and the bars are filled in orange.

```
ggplot(hof, aes(x = OPS)) +
  geom_histogram(
    breaks = seq(0.4, 1.2, by = 0.1),
    color = "white", fill = "orange"
  )
```

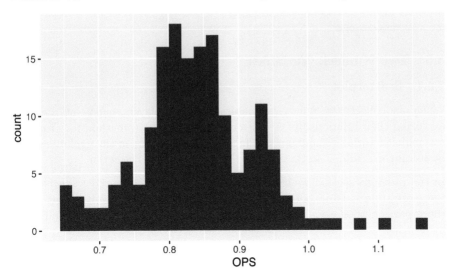

FIGURE 3.5
Histogram of the OPS values of the Hall of Fame players.

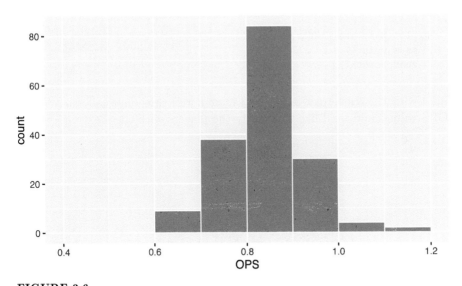

FIGURE 3.6
Histogram of the OPS values of the Hall of Fame players using different bins
and different color and fill options.

3.5 Two Numeric Variables

3.5.1 Scatterplot

When one collects two numeric variables for many players, one is interested in exploring their relationship. A scatterplot is a standard method for graphing two numeric variables, and one can produce a scatterplot in the **ggplot2** system by using the x and y aesthetics and the `geom_point()` function.

In the previous section we explored the distribution of the OPS statistic. Is there any relationship between a player's OPS and the baseball era? Were there particular seasons where the Hall of Fame OPS values were unusually high or low?

We can answer these questions by constructing a scatterplot using `geom_point()` where the variables `MidCareer` and `OPS` are respectively mapped to the x and y aesthetics. As it can be difficult to visually detect scatterplot patterns, it is helpful to add a smoothing curve by use of the `geom_smooth()` function to show the general association. This function by default implements the popular LOESS smoothing method (Cleveland 1979).

```
ggplot(hof, aes(MidCareer, OPS)) +
  geom_point() +
  geom_smooth()
```

In viewing the scatterplot in Figure 3.7, we notice three unusually large career OPS values, and we'd like to identify the players with these extreme values.

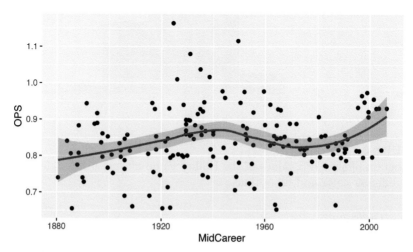

FIGURE 3.7
Scatterplot of the OPS and Midcareer values of the Hall of Fame players.

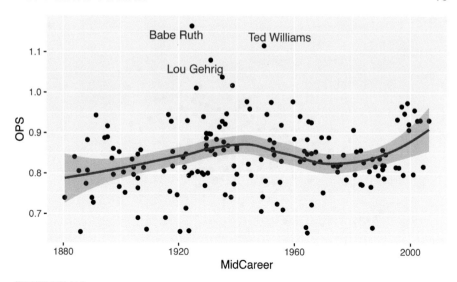

FIGURE 3.8
Scatterplot of the OPS and Midcareer values of the Hall of Fame players with
points identified.

Figure 3.8 shows the scatterplot with points identified. We achieve this by
adding text labels to the plot using the `geom_text_repel()` function form the
ggrepel package. Note that we use `filter()` to only send a small subset of
the data to this function. Also the labels are colored red by use of the `color
= "red"` argument to `geom_text_repel()`.

```
library(ggrepel)
ggplot(hof, aes(MidCareer, OPS)) +
  geom_point() +
  geom_smooth() +
  geom_text_repel(
    data = filter(hof, OPS > 1.05 | OPS < .5),
    aes(MidCareer, OPS, label = Player), color = "red"
  )
```

What do we learn from Figures 3.7 and 3.8? The typical OPS of a Hall of
Famer has stayed pretty constant through the years. But there was an increase
in the OPS during the 1930s when Babe Ruth and Lou Gehrig were in their
primes. It is interesting to note that the variability of the OPS values among
these players seems smaller in recent seasons.

3.5.2 Building a graph, step-by-step

Generally, constructing a graph is an iterative process. One begins by choosing
variables of interest and a particular graphical method (such as a scatterplot.

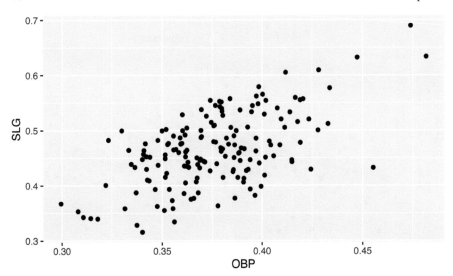

FIGURE 3.9
Scatterplot of the OPS and SLG values of the Hall of Fame players.

By inspecting the resulting display, one will typically find ways for the graph to be improved. By using several of the optional arguments, one can make changes to the graph that result in a clearer and more informative display. We illustrate this graph construction process in the situation where one is exploring the relationship between two variables.

There are two dimensions of hitting, the ability to get on base, measured by on-base percentage (OBP), and the ability to advance runners already on base, measured by slugging percentage (SLG). One can better understand the hitting performances of players by constructing a scatterplot of these two measures. We use the `geom_plot()` function to construct a scatterplot of OBP and SLG (see Figure 3.9).

```
(p <- ggplot(hof, aes(OBP, SLG)) + geom_point())
```

Looking at Figure 3.9, we see several problems with this display. Notably, the graph would be easier to read if more descriptive labels were used for the two axes. We plot a new figure to incorporate these new ideas. We use the `xlab()` and `ylab()` functions to replace OBP and SLG respectively with "On-Base Percentage" and "Slugging Percentage". The updated display is shown in Figure 3.10.

```
(p <- p +
   xlab("On Base Percentage") +
```

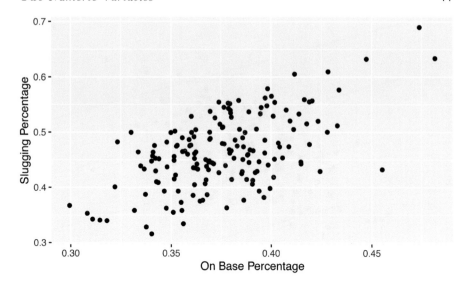

FIGURE 3.10
Scatterplot of the OPS and SLG values of the Hall of Fame players with descriptive labels for the two axes.

```
ylab("Slugging Percentage"))
```

Equivalently, we could change the limits and the labels by appealing directly to the `scale_x_continuous()` and `scale_y_continuous()` functions.

A good measure of batting performance is the OPS statistic defined by $OPS = OBP + SLG$. To evaluate hitters in our graph on the basis of OPS, it would be helpful to draw constant values of OPS on the graph. If we represent OBP and SLG by x and y, suppose we wish to draw a line where $OPS = 0.7$ or where $x + y = 0.7$. Equivalently, we want to draw the function $y = 0.7 - x$ on the graph; this is accomplished in the **ggplot2** system by the `geom_abline()` function where the arguments to the function are given by `slope` $= -1$ and `intercept` $= 0.7$. Similarly, we apply the `geom_abline()` function three more times to draw lines on the graph where OPS takes on the values 0.8, 0.9, and 1.0. The resulting display is shown in Figure 3.11.

```
(p <- p +
   geom_abline(
     slope = -1,
     intercept = seq(0.7, 1, by = 0.1),
     color = "red"
   )
)
```

In our final iteration, we add labels to the lines showing the constant values of OPS, and we label the points corresponding to players having a lifetime OPS

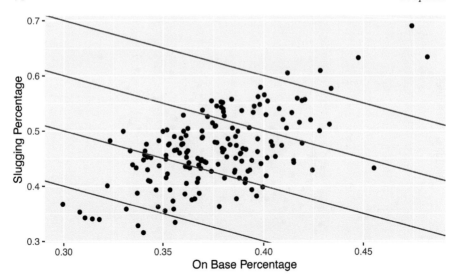

FIGURE 3.11
Scatterplot of the OPS and SLG values of the Hall of Fame players with reference lines.

exceeding one. Each of the line labels is accomplished using the `annotate()` function—the three arguments are the x location and y location where the text is to be drawn, and `label` is the vector of strings of text to be displayed (see Figure 3.12).

```
p +
  annotate(
    "text", angle = -13,
    x = rep(0.31, 4) ,
    y = seq(0.4, 0.7, by = 0.1) + 0.02,
    label = paste("OPS = ", seq(0.7, 1, by = 0.1)),
    color = "red"
  )
```

Rather than input these labels manually, we could create a data frame with the coordinates and labels, and then use the `geom_text()` function to add the labels to the plot.

```
ops_labels <- tibble(
  OBP = rep(0.3, 4),
  SLG = seq(0.4, 0.7, by = 0.1) + 0.02,
  label = paste("OPS =", OBP + SLG),
  angle = -13
)
```

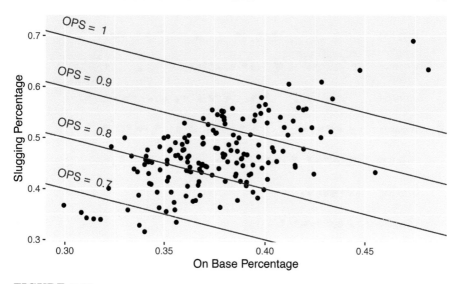

FIGURE 3.12
Scatterplot of the OPS and SLG values of the Hall of Fame players with reference lines and labels.

```
p +
  geom_text(
    data = ops_labels,
    hjust = "left",
    aes(label = label, angle = angle),
    color = "red"
  )
```

This final graph is very informative about the batting performance of these Hall of Famers. We see that a large group of these batters have career OPS values between 0.8 and 0.9, and only six players (Hank Greenberg, Rogers Hornsby, Jimmie Foxx, Ted Williams, Lou Gehrig, and Babe Ruth) had career OPS values exceeding 1.0. Points to the right of the major point cloud correspond to players with strong skills in getting on-base, but relatively weak in advancing runners home. In contrast the points to the left of the major point cloud correspond to hitters who are better in slugging than in reaching base.

3.6 A Numeric Variable and a Factor Variable

When one collects a numeric variable such as OPS and a factor such as era, one is typically interested in comparing the distributions of the numeric variable across different values of the factor. In the **ggplot2** system, the `geom_jitter()`

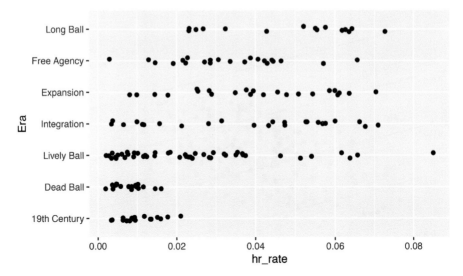

FIGURE 3.13
One-dimensional scatterplots of HR Rates by era.

function can be used to construct parallel stripcharts or number line graphs
for values of the factor, and the `geom_boxplot()` function constructs parallel
boxplots (graphs of summaries of the numeric variable) across the factor.

Home run hitting has gone through dramatic changes in the history of baseball,
and suppose we are interested in exploring these changes over baseball eras.
Suppose one focuses on the home run rate defined by HR/AB for our Hall of
Fame players. We add a new variable `hr_rate` to the data frame `hof`:

```
hof <- hof |>
   mutate(hr_rate = HR / AB)
```

3.6.1   Parallel stripcharts

One constructs parallel stripcharts of `hr_rate` by `Era` by using the
`geom_jitter()` function; the x and y aesthetics are mapped to `hr_rate` and
`Era`, respectively. We use the `height = 0.1` argument to reduce the amount
of the vertical jitter of the points.

```
ggplot(hof, aes(hr_rate, Era)) +
   geom_jitter(height = 0.1)
```

Figure 3.13 shows how the rate of hitting home runs has changed over eras.
Home runs were rare in the 19th Century and Dead Ball eras. In the Lively
Ball era, home run hitting was still relatively low, but there were some

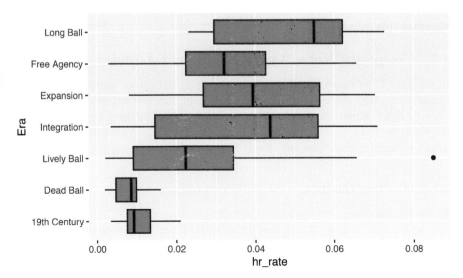

FIGURE 3.14
Parallel boxplots of HR Rates by era.

unusually good home run hitters such as Babe Ruth. The home run rates in the Integration, Expansion, and Free Agency eras were pretty similar.

3.6.2 Parallel boxplots

An alternative display for comparing distributions uses the `geom_boxplot()` function. Here the `x` and `y` aesthetics are mapped to `Era` and `hr_rate`, respectively. The function `coord_flip()` will flip the axes and display the boxplots horizontally. By use of the `color` and `fill` arguments, we display orange boxplots with brown edges.

```
ggplot(hof, aes(Era, hr_rate)) +
  geom_boxplot(color = "brown", fill = "orange") +
  coord_flip()
```

The parallel boxplot display is shown in Figure 3.14. Each rectangle in the display shows the location of the lower quartile, the median, and the upper quartile, and lines are drawn to the extreme values. Unusual points (outliers) that fall far from the rest of the distribution are indicated by points outside the boxes. This graph confirms the observations we made when we viewed the stripchart display. Home run hitting was low in the first two eras and started to increase in the Lively Ball era. It is interesting that the only outlier among these Hall of Famers was Babe Ruth's career home run rate of 0.085.

3.7 Comparing Ruth, Aaron, Bonds, and A-Rod

In Chapter 1, we constructed a graph comparing the career home run trajectories of four great sluggers in baseball history. In this section, we describe how we used R to create this graph. First, we need to load in the relevant data into R. Next, we need to construct data frames containing the home run and age data for the sluggers. Last, we use R functions to construct the graph.

3.7.1 Getting the data

To obtain the graph, we need to collect the number of home runs, at-bats, and the age for each season of each slugger's career. From the **Lahman** package, the relevant data frames are `People` and `Batting`. From the data frame `People`, we obtain the player ids and birth years for the four players. The `Batting` data frame is used to extract the home run and at-bats information.

We begin by reading in the **Lahman** package.

```
library(Lahman)
```

From the `People` data frame, we wish to extract the player id and the birth year for a particular player.

- The `filter()` function is used to extract the rows in the `People` data frame matching each player's id.
- In Major League Baseball, a player's age for a season is defined to be his age on June 30. So we make a slight adjustment to a player's birth year depending if his birthday falls in the first six months or not. The adjusted birth year is stored in the variable `mlb_birthyear`. (The `if_else()` function is useful for assignments based on a condition; if `birthMonth >= 7` is TRUE, then `birthyear <- birthYear + 1`, otherwise `birthyear <- birthyear`.)

```
PlayerInfo <- People |>
  filter(
    playerID %in% c(
      "ruthba01", "aaronha01", "bondsba01", "rodrial01"
    )
  ) |>
  mutate(
    mlb_birthyear = if_else(
      birthMonth >= 7, birthYear + 1, birthYear
    ),
```

```
   Player = paste(nameFirst, nameLast)
 ) |>
 select(playerID, Player, mlb_birthyear)
```

The `PlayerInfo` data frame contains information for the sluggers Babe Ruth, Hank Aaron, Barry Bonds, and Alex Rodriguez.

3.7.2 Creating the player data frames

Now that we have the player id codes and birth years, we use this information together with the **Lahman** batting data frame `Batting` to create data frames for each of these four players. One of the variables in the batting data frame is `playerID`. To get the batting and age data for Babe Ruth, we use the `inner_join()` function to match the rows of the batting data to those corresponding in the `PlayerInfo` data frame where `playerID` is equal. We create a new variable `Age` defined to be the season year minus the player's birth year. (Recall that we made a slight modification to the `birthyear` variable so that one obtains a player's correct age for a season.) Last, for each player, we use the `cumsum()` function on the grouped data to create a new variable `cHR` containing the cumulative count of home runs for each player each season.

```
HR_data <- Batting |>
  inner_join(PlayerInfo, by = "playerID") |>
  mutate(Age = yearID - mlb_birthyear) |>
  select(Player, Age, HR) |>
  group_by(Player) |>
  mutate(cHR = cumsum(HR))
```

3.7.3 Constructing the graph

We want to plot the cumulative home run counts for each of the four players against age. In the data frame `HR_data` the relevant variables are `cHR`, `Age`, and `Player`. We use the `geom_line()` function to graph the cumulative home run counts against age. By mapping the `color` aesthetic to the `Player` variable, distinct cumulative home run lines are drawn for each player. Note that different colors are used for the four players and a legend is automatically constructed that matches up the line type with the player's name. The **scale_color_manual** function allows us to specify the set of colors to use in the plot. In this case, the vector `crc_fc` contains an ordered set of pre-defined colors.

```
ggplot(HR_data, aes(x = Age, y = cHR, color = Player)) +
  geom_line() +
  scale_color_manual(values = crc_fc)
```

Figure 3.15 displays the completed graph.

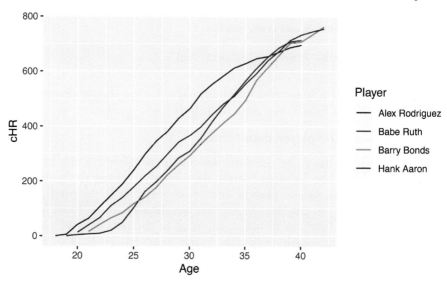

FIGURE 3.15
Cumulative home run counts against age for four ballplayers.

3.8 The 1998 Home Run Race

The Retrosheet play-by-play files are helpful for learning about patterns of player performance during a particular baseball season. We illustrate the use of R to read in the files for the 1998 season and graphically view the famous home run duel between Mark McGwire and Sammy Sosa.

3.8.1 Getting the data

We begin by reading in the 1998 play-by-play data and storing it in the data frame `retro1998`. See Section A.1.3 for information about how to create this file.

```
retro1998 <- read_rds(here::here("data/retro1998.rds"))
```

In the play-by-play data, the variable `bat_id` gives the identification code for the player who is batting. To extract the batting data for McGwire and Sosa, we need to find the codes for these two players available in the **Lahman People** data frame. By use of the `filter()` function, we find the id code where `nameFirst = "Sammy"` and `nameLast = "Sosa"`. Likewise, we find the id code corresponding to Mark McGwire; these codes are stored in the variables `sosa_id` and `mac_id`.

```
sosa_id <- People |>
  filter(nameFirst == "Sammy", nameLast == "Sosa") |>
  pull(retroID)
mac_id <- People |>
  filter(nameFirst == "Mark", nameLast == "McGwire") |>
  pull(retroID)
```

Now that we have the player id codes, we extract McGwire's and Sosa's plate appearance data from the play-by-play data frame `retro1998`. These data are stored in the data frame `hr_race`.

```
hr_race <- retro1998 |>
  filter(bat_id %in% c(sosa_id, mac_id))
```

3.8.2 Extracting the variables

For each player, we are interested in collecting the current number of home runs hit for each plate appearance and graphing the date against the home run count. For each player, the two important variables are the date and the home run count. We write a new function `cum_hr()` that will extract these two variables given a player's play-by-play batting data.

In the play-by-play data frame, the variable `game_id` identifies the game location and date. For example, the value `game_id` of `ARI199805110` indicates that this particular play occurred at the game played in Arizona on May 11, 1998. (The variable is displayed in the "location, year, month, day" format.) Using the `str_sub()` function, we select the 4th through 11th characters of this string variable and assign this date to the variable `Date`. (The `ymd()` function converts the date to the more readable "year-month-day" format, and forces R to recognize it as a `Date`.) Using the `arrange()` function, we sort the play-by-play data from the beginning to the end of the season. The variable `event_cd` contains the outcome of the batting play; a value `event_cd` of 23 indicates that a home run has been hit. We define a new variable `HR` to be either 1 or 0 depending if a home run occurred, and the new variable `cumHR` records the cumulative number of home runs hit in the season using the `cumsum()` function. The output of the function is a new data frame containing each date and the cumulative number of home runs to date for all plate appearances during the season.

```
cum_hr <- function(data) {
  data |>
    mutate(Date = ymd(str_sub(game_id, 4, 11))) |>
    arrange(Date) |>
    mutate(
```

```
      HR = if_else(event_cd == 23, 1, 0),
      cumHR = cumsum(HR)
    ) |>
    select(Date, cumHR)
}
```

After grouping the `hr_race` data frame by player, and collecting the corresponding player ids, we use the `group_split()` and `map()` functions to iterate `cum_hr()` twice, once on Sosa's batting data and once on McGwire's batting data, obtaining the new data frame `hr_ytd`.

```
hr_grouped <- hr_race |>
  group_by(bat_id)

keys <- hr_grouped |>
  group_keys() |>
  pull(bat_id)

hr_ytd <- hr_grouped |>
  group_split() |>
  map(cum_hr) |>
  set_names(keys) |>
  bind_rows(.id = "bat_id") |>
  inner_join(People, by = c("bat_id" = "retroID"))
```

3.8.3 Constructing the graph

Once this new data frame is created, it is straightforward to produce the graph of interest. The `geom_line()` function constructs a graph of the cumulative home run count against the date. By mapping `nameLast` to the `color` aesthetic, the lines corresponding to the two players are drawn using different colors. We use the `geom_hline()` function to add a horizontal line at the home run value of 62 and the `annotate()` function is applied to place the text string "62" above this plotted line (see Figure 3.16).

```
ggplot(hr_ytd, aes(Date, cumHR, color = nameLast)) +
  geom_line() +
  geom_hline(yintercept = 62, color = crcblue) +
  annotate(
    "text", ymd("1998-04-15"), 65,
    label = "62", color = crcblue
  ) +
  ylab("Home Runs in the Season")
```

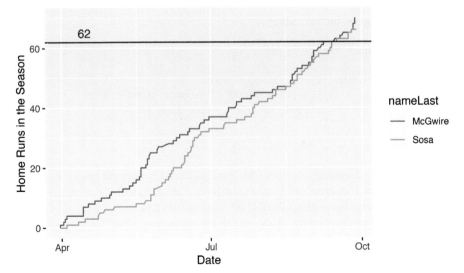

FIGURE 3.16
Graph of the 1998 home run race between Sammy Sosa and Mark McGwire.

3.9 Further Reading

A good reference to the traditional graphics system in R is Murrell (2006). Kabacoff (2010) together with the Quick-R website at https://www.statmeth ods.net provide a useful reference for specific graphics functions. Chapter 4 of Albert and Rizzo (2012) provides a number of examples of modifying traditional graphics in R such as changing the plot type and symbol, using color, and overlying curves and mathematical expressions. Wickham, Çetinkaya-Rundel, and Grolemund (2023), Benjamin S. Baumer, Kaplan, and Horton (2021b) and Ismay and Kim (2019) all discuss the use of **ggplot2** for creating data graphics.

3.10 Exercises

1. Hall of Fame Pitching Dataset

The `hof_pitching` data frame in the **abdwr3edata** package contains the career pitching statistics for all of the pitchers inducted in the Hall of Fame. The variable BF is the number of batters faced by a pitcher in his career. Suppose we group the pitchers by this variable using the intervals (0, 10,000), (10,000, 15,000), (15,000, 20,000), (20,000, 30,000). One can reexpress the variable BF to the grouped variable BF_group by use of the `cut()` function.

```
hofpitching <- hofpitching |>
  mutate(
    BF_group = cut(
      BF,
      c(0, 10000, 15000, 20000, 30000),
      labels = c("Less than 10000", "(10000, 15000)",
                 "(15000, 20000)", "more than 20000")
    )
  )
```

a. Construct a frequency table of BF.group using the summarize() function.
b. Construct a bar graph of the output from summarize(). How many HOF pitchers faced more than 20,000 pitchers in their career?
c. Construct an alternative graph of the BF.group variable. Compare the effectiveness of the bar graph and the new graph in comparing the frequencies in the four intervals.

2. Hall of Fame Pitching Dataset (Continued)

The variable WAR is the total wins above replacement of the pitcher during his career.

a. Using the geom_histogram() function, construct a histogram of WAR for the pitchers in the Hall of Fame dataset.

b. There are two pitchers who stand out among all of the Hall of Famers on the total WAR variable. Identify these two pitchers.

3. Hall of Fame Pitching Dataset (Continued)

To understand a pitcher's season contribution, suppose we define the new variable WAR_Season defined by

```
hofpitching <- hofpitching |>
  mutate(WAR_Season = WAR / Yrs)
```

a. Use the geom_point() function to construct parallel one-dimensional scatterplots of WAR.Season for the different levels of BP.group.
b. Use the geom_boxplot() function to construct parallel boxplots of WAR.Season across BP.group.
c. Based on your graphs, how does the wins above replacement per season depend on the number of batters faced?

4. Hall of Fame Pitching Dataset (Continued)

Suppose we limit our exploration to pitchers whose mid-career was 1960 or later. We first define the MidYear variable and then use the filter() function to construct a data frame consisting of only these 1960+ pitchers.

```
hofpitching <- hofpitching |>
   mutate(MidYear = (From + To) / 2)
hofpitching.recent <- hofpitching |>
   filter(MidYear >= 1960)
```

a. By use of the `arrange()` function, order the rows of the data frame by the value of `WAR_Season`.
b. Construct a dot plot of the values of `WAR_Season` where the labels are the pitcher names.
c. Which two 1960+ pitchers stand out with respect to wins above replacement per season?

5. Hall of Fame Pitching Dataset (Continued)

The variables `MidYear` and `WAR_Season` are defined in the previous exercises.

a. Construct a scatterplot of `MidYear` (horizontal) against `WAR_Season` (vertical).
b. Is there a general pattern in this scatterplot? Explain.
c. There are two pitchers whose mid careers were in the 1800s who had relatively low `WAR_Season` values. By use of the `filter()` and `geom_text()` functions, add the names of these two pitchers to the scatterplot.

6. Working with the Lahman Batting Dataset

a. Read the **Lahman** `People` and `Batting` data frames into R.
b. Collect in a single data frame the season batting statistics for the great hitters Ty Cobb, Ted Williams, and Pete Rose.
c. Add the variable `Age` to each data frame corresponding to the ages of the three players.
d. Using the `geom_line()` function, construct a line graph of the cumulative hit totals against age for Pete Rose.
e. Using the `geom_line()` function, overlay the cumulative hit totals for Cobb and Williams.
f. Write a short paragraph summarizing what you have learned about the hitting pattern of these three players.

7. Working with the Lahman Teams Dataset

The **Lahman** `Teams` dataset contains yearly statistics and standing information for all teams in MLB history.

a. Read the `Teams` data frame into R.
b. Create a new variable `win_pct` defined to be the team winning percentage `W / (W + L)`.
c. For the teams in the 2022 season, construct a scatterplot of the team ERA and its winning percentage.
d. Use the `geom_mlb_scoreboard_logos()` function from the **mlbplotR** package to put the team logos on the scatterplot as plotting marks.

Use this function to redo the graph in part (c), plotting using the team logos.

8. Working with the Retrosheet Play-by-Play Dataset

In Section 3.8, we used the Retrosheet play-by-play data to explore the home run race between Mark McGwire and Sammy Sosa in the 1998 season. Another way to compare the patterns of home run hitting of the two players is to compute the spacings, the number of plate appearances between home runs.

a. Following the work in Section 3.8, create the two data frames `mac_data` and `sosa_data` containing the batting data for the two players.

b. Use the following R commands to restrict the two data frames to the plays where a batting event occurred. (The relevant variable `bat_event_fl` is either `TRUE` or `FALSE`.)

```
mac_data <- filter(mac_data, bat_event_fl == TRUE)
sosa_data <- filter(sosa_data, bat_event_fl == TRUE)
```

c. For each data frame, create a new variable `PA` that numbers the plate appearances 1, 2, ... (The function `nrow()` gives the number of rows of a data frame.)

```
mac_data <- mutate(mac_data, PA = 1:nrow(.))
sosa_data <- mutate(sosa_data, PA = 1:nrow(.))
```

d. The following commands will return the numbers of the plate appearances when the players hit home runs.

```
mac_HRPA <- mac.data |>
  filter(event_cd == 23) |>
  pull(PA)
sosa_HRPA <- sosa.data |>
  filter(event_cd == 23) |>
  pull(PA)
```

e. Using the R function `diff()`, the following commands compute the spacings between the occurrences of home runs.

```
mac_spacings <- diff(c(0, mac_HRPA))
sosa_spacings <- diff(c(0, sosa_HRPA))
```

Create a new data frame `HR_Spacing` with two variables, `Player`, the player name, and `Spacing`, the value of the spacing. f. By use of the `summarize()` and `geom_histogram()` functions on the data frame `HR_Spacing`, compare the home run spacings of the two players.

4

The Relation Between Runs and Wins

4.1 Introduction

The goal of a baseball team is—just like a team in any other sport—to win games. Similarly, the goal of the baseball analyst is being able to measure what happens on the field in term of wins. Answering a question such as "Who is the better player between Dee Gordon and J.D. Martinez?" becomes an easier task if one succeeds in estimating how much Gordon's speed and slick fielding contribute to his team's victories and how many wins can be attributed to Martinez's powerful bat.

Victories are obtained by outscoring opponents, thus the percentage of wins obtained by a team over the course of a season is strongly correlated with the number of runs it scores and allows. This chapter explores the relationship between runs and wins. Understanding this relationship is a critical step toward answering questions about players' value. In fact, while it's impossible to directly quantify the impact of players in terms of wins, we will show in the following chapters that it is possible to estimate their contributions in term of runs.

4.2 The Teams Table in the Lahman Database

The `Teams` table from the **Lahman** package contains seasonal stats for major league teams going back to the first professional season in 1871. We begin by loading these data into R and exploring their contents by looking at the final lines of this dataset, using the `slice_tail()` function.

```
library(Lahman)
Teams |>
```

DOI: 10.1201/9781032668239-4

```
slice_tail(n = 3)
```

	yearID	lgID	teamID	franchID	divID	Rank	G	Ghome	W	L
1	2022	AL	TEX	TEX	W	4	162	81	68	94
2	2022	AL	TOR	TOR	E	2	162	81	92	70
3	2022	NL	WAS	WSN	E	5	162	81	55	107

	DivWin	WCWin	LgWin	WSWin	R	AB	H	X2B	X3B	HR	BB	SO
1	N	N	N	N	707	5478	1308	224	20	198	456	1446
2	N	Y	N	N	775	5555	1464	307	12	200	500	1242
3	N	N	N	N	603	5434	1351	252	20	136	442	1221

	SB	CS	HBP	SF	RA	ER	ERA	CG	SHO	SV	IPouts	HA	HRA	BBA	SOA
1	128	41	47	38	743	673	4.22	1	10	37	4305	1345	169	581	1314
2	67	35	55	33	679	620	3.87	0	10	46	4324	1356	180	424	1390
3	75	31	60	37	855	785	5.00	2	4	28	4235	1469	244	558	1220

	E	DP	FP	name	park
1	96	143	0.984	Texas Rangers	Globe Life Field
2	82	120	0.986	Toronto Blue Jays	Rogers Centre
3	104	126	0.982	Washington Nationals	Nationals Park

	attendance	BPF	PPF	teamIDBR	teamIDlahman45	teamIDretro
1	2011361	100	101	TEX	TEX	TEX
2	2653830	100	100	TOR	TOR	TOR
3	2026401	94	96	WSN	MON	WAS

The description of every column is provided in the help files accompanying the **Lahman** package (e.g. help(Teams)).

Suppose that one is interested in relating the proportion of wins with the runs scored and runs allowed for all of the teams. Toward this goal, the relevant fields of interest in this table are the number of games played G, the number of team wins W, the number of losses L, the total number of runs scored R, and the total number of runs allowed RA. We create a new data frame my_teams containing only the above five columns plus information on the team (teamID), the season (yearID), and the league (lgID). We are interested in studying the relationship between wins and runs for recent seasons, so we use the filter() function to focus our exploration on seasons since 2001.

```
my_teams <- Teams |>
  filter(yearID > 2000) |>
  select(teamID, yearID, lgID, G, W, L, R, RA)
my_teams |>
  slice_tail(n = 6)
```

	teamID	yearID	lgID	G	W	L	R	RA
1	SFN	2022	NL	162	81	81	716	697
2	SLN	2022	NL	162	93	69	772	637

3	TBA	2022	AL	162	86	76	666	614
4	TEX	2022	AL	162	68	94	707	743
5	TOR	2022	AL	162	92	70	775	679
6	WAS	2022	NL	162	55	107	603	855

The run differential is defined as the difference between the runs scored and the runs allowed by a team. The winning proportion is the fraction of games won by a team. In baseball (and generally in sports) *winning percentage* is commonly used instead of the more appropriate winning proportion. In the remainder of this chapter we have chosen to adopt the most widely used term. We calculate two new variables RD (run differential) and Wpct (winning percentage) with the following lines of code.

```
my_teams <- my_teams |>
  mutate(RD = R - RA, Wpct = W / (W + L))
```

A scatterplot of the run differential and the winning percentage gives a first indication of the association between the two variables. Here we create the plot and store it as **run_diff**. We delay its appearance as we will subsequently add to it.

```
run_diff <- ggplot(my_teams, aes(x = RD, y = Wpct)) +
  geom_point() +
  scale_x_continuous("Run differential") +
  scale_y_continuous("Winning percentage")
```

4.3 Linear Regression

One simple way to predict a team's winning percentage using runs scored and allowed is with linear regression. A simple linear model is

$$Wpct = a + b \times RD + \epsilon,$$

where a and b are unknown constants and ϵ is the error term that captures all other factors influencing the response variable (Wpct). This is a special case of a linear model fit using the lm() function from the **stats** package (which is installed and loaded in R by default). The most basic call to the function requires a formula, specified as **response ~ predictor1 + predictor2 + ..., data = dataset**, in which the variable to be modeled (a.k.a. the *dependent variable*) is indicated on the left side of the tilde character (~) and the variables used to predict the response are specified on the right side. In the following illustration of the lm() function, we use the **data** argument to specify which data frame to use.

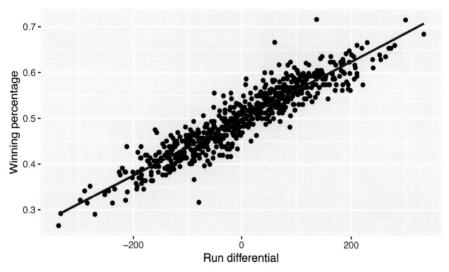

FIGURE 4.1
Scatterplot of team run differential against team winning percentage for major
league teams from 2001 to 2022. A best-fitting line is overlaid on top of the
scatterplot.

```
linfit <- lm(Wpct ~ RD, data = my_teams)
linfit
```

```
Call:
lm(formula = Wpct ~ RD, data = my_teams)

Coefficients:
(Intercept)              RD
   0.499985        0.000624
```

When executing the code `run_diff`, a scatterplot is displayed in Figure 4.1
that shows a strong positive relationship—teams with large run differentials
are more likely to be winning. The fitted line in the plot of Figure 4.1 is drawn
with the `geom_smooth()` command, which matches the output of `lm()` when
the `method` argument is set to `"lm"`.

```
run_diff +
   geom_smooth(method = "lm", se = FALSE, color = crcblue)
```

From the above output, a team's expected winning percentage can be estimated
from its run differential RD by the equation:

$$\widehat{Wpct} = 0.499985 + 0.000624 \times RD \qquad (4.1)$$

This formula tells us that a team with a run differential of zero ($RD = 0$) should win half of its games (estimated intercept $\approx .500$)—a desired property. In addition, a one-unit increase in run differential corresponds to an increase of 0.000624 in winning percentage. To give further insight into this relationship, a team scoring 750 runs and allowing 750 runs is predicted to win half of its games corresponding to 81 games in a typical MLB season of 162 games. In contrast, a team scoring 760 runs and allowing 750 has a run differential of +10 and is predicted to have a winning percentage of $0.500 + 10 \cdot 0.000624 \approx 0.506$. A winning percentage of 0.506 in a 162-game schedule corresponds to 82 wins. Thus an increase of 10 runs in the run differential of a team corresponds— according to the straight-line model—to an additional expected win in the standings.

One concern about this model is that predictions from this fitted line can fall outside the range $[0, 1]$. For example, a hypothetical team that outscores its opponent by a total of 805 runs would be predicted to win more than 100 percent of its games—which is impossible. However, since over 99 percent of teams throughout major league baseball history have run differentials between –350 and +350, the straight-line model is reasonable.

Once we have a fitted model, we use the function `augment()` from the **broom** package to calculate the predicted values from the model, as well as the residuals, which measure the difference between the response values and the fitted values (i.e., between the actual and the estimated winning percentages).

```
library(broom)
my_teams_aug <- augment(linfit, data = my_teams)
```

Figure 4.2 displays a plot of the residuals against the run differential. Note that the `.resid` variable was created by `augment()` and stores the residuals. We use the **ggrepel** package to label a few teams with the largest residuals.

```
base_plot <- ggplot(my_teams_aug, aes(x = RD, y = .resid)) +
  geom_point(alpha = 0.3) +
  geom_hline(yintercept = 0, linetype = 3) +
  xlab("Run differential") + ylab("Residual")

highlight_teams <- my_teams_aug |>
  arrange(desc(abs(.resid))) |>
  slice_head(n = 4)

library(ggrepel)
base_plot +
```

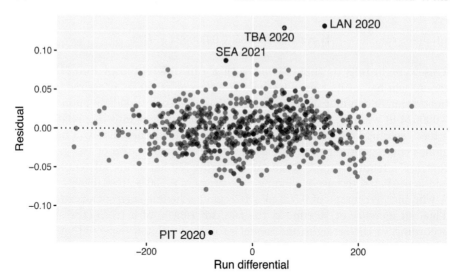

FIGURE 4.2
Residuals versus run differential for the fitted linear model. Four teams with
large residuals are labeled.

```
geom_point(data = highlight_teams, color = crcblue) +
geom_text_repel(
  data = highlight_teams, color = crcblue,
  aes(label = paste(teamID, yearID))
)
```

Residuals can be interpreted as the error of the linear model in predicting
the actual winning percentage. Thus the points in Figure 4.2 farthest from
the zero line correspond to the teams where the linear model fared worst in
predicting the winning percentage.

One of the extreme values at the top of the residual graph in Figure 4.2
corresponds to the 2008 Los Angeles Angels: given their +68 run differential,
they were supposed to have a 0.542 winning percentage according to the linear
equation (Equation 4.1). However, they ended the season at 0.617. The residual
value for this team is $0.617 - 0.542 = 0.075$, or $0.075 \cdot 162 = 12.2$ games. At
the other end of the spectrum, the 2006 Cleveland Indians, with a +88 run
differential, are seen as a 0.555 team by the linear model, but they actually
finished at a mere 0.481, corresponding to the residual $0.481 - 0.555 = -0.073$,
or -11.8 games.

The average value of the residuals for any least squares linear model is equal to
zero, which means that the model predictions are equally likely to overestimate
and underestimate the winning percentage. Statistically, we say that the method

for fitting the model is unbiased. In order to estimate the average magnitude of the errors, we first square the residuals so that each error has a positive value, calculate the mean of the squared residuals, and take the square root of each mean value to get back to the original scale. The value so calculated is the *root mean square error*, abbreviated as RMSE. (Note the use of the square root function `sqrt()`.)

```
resid_summary <- my_teams_aug |>
  summarize(
    N = n(),
    avg = mean(.resid),
    RMSE = sqrt(mean(.resid^2))
  )
resid_summary
```

```
# A tibble: 1 x 3
      N       avg    RMSE
  <int>     <dbl>   <dbl>
1   660  4.21e-16  0.0280
```

```
rmse <- resid_summary |>
  pull(RMSE)
```

If the errors are normally distributed, approximately two thirds of the residuals fall between $-RMSE$ and $+RMSE$, while 95% of the residuals are between $-2 \cdot RMSE$ and $2 \cdot RMSE$.[1] These statements can be confirmed with the following lines of code. (The function `abs()` computes the absolute value.)

```
my_teams_aug |>
  summarize(
    N = n(),
    within_one = sum(abs(.resid) < rmse),
    within_two = sum(abs(.resid) < 2 * rmse)
  ) |>
  mutate(
    within_one_pct = within_one / N,
    within_two_pct = within_two / N
  )
```

```
# A tibble: 1 x 5
      N within_one within_two within_one_pct within_two_pct
  <int>      <int>      <int>          <dbl>          <dbl>
```

---

[1] Equivalently, over a 162-game season the number of wins predicted by the linear model comes within four wins of the actual number of wins in two-thirds of the cases, while for 19 out of 20 teams the difference is not higher than 8 wins.}

| 1 | 660 | 476 | 629 | 0.721 | 0.953 |

We use the function `n()` in conjunction with `summarize()` above to obtain the number of rows of a data frame. In the numerators of the expressions, we obtain the number of residuals (computed using the `abs()` function) that are smaller than one and two $RMSE$. The computed fractions are close to the theoretical 68% and 95% values stated above.

4.4   The Pythagorean Formula for Winning Percentage

Bill James, regarded as the godfather of sabermetrics, empirically derived the following non-linear formula to estimate winning percentage, called the *Pythagorean expectation*.

$$\widehat{Wpct} = \frac{R^2}{R^2 + RA^2} \tag{4.2}$$

One can use this formula to predict winning percentages by use of the following R code.

```
my_teams <- my_teams |>
  mutate(Wpct_pyt = R ^ 2 / (R ^ 2 + RA ^ 2))
```

Here the residuals need to be calculated explicitly, but that's not a hard task. We define a new variable `residuals_pyt` that is the difference between the actual and predicted winning percentages. We also compare the RMSE for these new predictions.

```
my_teams <- my_teams |>
  mutate(residuals_pyt = Wpct - Wpct_pyt)
my_teams |>
  summarize(rmse = sqrt(mean(residuals_pyt^2)))
```

```
    rmse
1 0.0265
```

The RMSE calculated on the Pythagorean predictions is similar in value to the one calculated with the linear predictions (it's actually slightly lower for the 2000–2022 data we have been using here). Thus, if the complex non-linear model is not more accurate, why should we use it? In fact, the Pythagorean expectation model has several desirable properties missing in the linear model. Both of these advantages can be illustrated with simple examples.

Suppose there exists a powerhouse team that scores an average of ten runs per game, while allowing an average of close to five runs per game. In a 162-game schedule, this team would score 1620 runs, while allowing 810, for a run differential of 810. Replacing RD with 810 in the linear equation, one obtains a winning percentage of over 1, which is impossible. On the other hand, replacing R and RA with 1620 and 810 respectively in the Pythagorean expectation, the resulting winning percentage is equal to 0.8, a more reasonable prediction. Suppose a second hypothetical team has pitchers who never allow runs, while the hitters always manage to score the only run they need. Such a team will score 162 runs in a season and win all of its games, but the linear equation would predict it to be merely a .601 team. The Pythagorean model instead correctly predicts that this team will win all of its games.

While neither of the above examples is realistic, there are some extreme situations in modern baseball history that push the utility of the linear model. For example, the 2001 Seattle Mariners had 116 wins and 46 losses for a +300 run differential and the 2003 Detroit Tigers had a 43-119 record with a −337 run differential. In these unlikely scenarios, the Pythagorean model will give more sensible winning percentage estimates.

Finally, recall our statement at the end of the introductory section that the runs-to-wins relationship is crucial in assessing the contribution of players to their team's wins. Once we estimate the number of runs players contribute to their teams (as it will be shown in the following chapters), runs-to-wins formulas can be used to convert these run values to wins. One can now answer questions like "How many wins would a lineup of nine Mike Trouts accumulate in a season?" For these kinds of investigations, the scenarios in which the linear formula break down are more likely to occur, thus highlighting the need for a formula such as the Pythagorean expectation that gives reasonable predictions in all cases.

4.4.1   The Exponent in the Pythagorean model

Subsequent refinements to the Pythagorean model by Bill James and other analysts have aimed at finding an exponent that would give a better fit relative to the originally proposed exponent value of 2. In this section, we describe how one finds the value of the Pythagorean exponent leading to predictions closest to the actual winning percentages.

Replacing the value 2 in the Equation 4.2 with an unknown variable k, we write the formula as:

$$\frac{W}{W + L} = Wpct \approx \widehat{Wpct} = \frac{R^k}{R^k + RA^k}. \tag{4.3}$$

With some algebra, this equation can be rewritten as follows:

$$\frac{W}{L} \approx \frac{R^k}{RA^k}. \tag{4.4}$$

Taking the logarithm on both sides of the equation, we obtain the linear relationship

$$log\left(\frac{W}{L}\right) \approx k \cdot log\left(\frac{R}{RA}\right). \tag{4.5}$$

The value of k can now be estimated using linear regression, where the response variable is `log(W/L)` and the predictor is `log(R/RA)`. In the following R code, we compute the logarithm of the ratio of wins to losses, the logarithm of the ratio of runs to runs allowed, and fit a simple linear model with these transformed variables. (In the call to the `lm()` function, we specify a model with a zero intercept by adding a zero term on the right side of the formula.)

```
my_teams <- my_teams |>
  mutate(
    logWratio = log(W / L),
    logRratio = log(R / RA)
  )

pytFit <- lm(logWratio ~ 0 + logRratio, data = my_teams)
pytFit
```

```
Call:
lm(formula = logWratio ~ 0 + logRratio, data = my_teams)

Coefficients:
logRratio
     1.83
```

The R output suggests a best-fit Pythagorean exponent of 1.83, which is notably smaller than the value 2.

4.4.2 Good and bad predictions by the Pythagorean model

The 2011 Boston Red Sox scored 875 runs, while allowing 737. According to the Pythagorean model with exponent 2, they were expected to win 95 games—we obtain this number by plugging 875 and 737 into the Pythagorean formula and multiplying by the number of games in a season:

$$162 \times \frac{875^2}{875^2 + 737^2} \approx 95.$$

The Red Sox actually won 90 games. The five game difference was quite costly to the Red Sox, as they missed clinching the Wild Card (which went to the Tampa Bay Rays in the final game (actually in the final minute) of the regular season. The Pythagorean model is more on target with the Rays of the same season, as the prediction of 92 (coming from their 707 runs scored versus 614 runs allowed) is just a bit higher than the actual 91.

Why does the Pythagorean formula miss so poorly on the Red Sox? In other words, why did they win five fewer games than expected from their run differential? Let's have a look at their season game by game.

The data frame `retro_gl_2011` (a game log file downloaded from Retrosheet, see Section 1.3.3) contains detailed information on every game played in the 2011 season. The following commands load the file into R, select the lines pertaining to the Red Sox games, and keep only the columns related to runs.

```
library(abdwr3edata)
gl2011 <- retro_gl_2011

BOS2011 <- gl2011 |>
  filter(HomeTeam == "BOS" | VisitingTeam == "BOS") |>
  select(
    VisitingTeam, HomeTeam,
    VisitorRunsScored, HomeRunsScore
  )
slice_head(BOS2011, n = 6)
```

```
# A tibble: 6 x 4
  VisitingTeam HomeTeam VisitorRunsScored HomeRunsScore
  <chr>        <chr>                <dbl>         <dbl>
1 BOS          TEX                      5             9
2 BOS          TEX                      5            12
3 BOS          TEX                      1             5
4 BOS          CLE                      1             3
5 BOS          CLE                      4             8
6 BOS          CLE                      0             1
```

Using the results of every game featuring the Boston team, we calculate run differentials (`ScoreDiff`) both for games won and lost and add a column `W` indicating whether the Red Sox won the game.

```
BOS2011 <- BOS2011 |>
  mutate(
    ScoreDiff = ifelse(
      HomeTeam == "BOS",
      HomeRunsScore - VisitorRunsScored,
      VisitorRunsScored - HomeRunsScore
    ),
    W = ScoreDiff > 0
  )
```

We compute summary statistics on the run differentials for games won and for games lost using the `skim()` function from the **skimr** package, in conjunction with `group_by()`. To `group_by()`, we specify a grouping factor (i.e., whether the game resulted in a win for Boston). The `skim()` function takes a variable name and computes a host of relevant summary statistics, including the mean, standard deviation, and number of cases.

```
library(skimr)
BOS2011 |>
  group_by(W) |>
  skim(ScoreDiff) |>
  print(include_summary = FALSE)
```

```
-- Variable type: numeric ---------------------------------------
  skim_variable W      n_missing complete_rate  mean   sd  p0 p25
1 ScoreDiff     FALSE         0            1  -3.46 2.56 -11  -4
2 ScoreDiff     TRUE          0            1   4.3  3.28   1   2
  p50 p75 p100 hist
1  -3  -1   -1
2   4   6   14
```

The 2011 Red Sox had their victories decided by a larger margin than their losses (4.3 vs –3.5 runs on average), leading to their underperformance of the Pythagorean prediction by five games. A team overperforming (or underperforming) its Pythagorean winning percentage is often seen, in sabermetrics circles, as being lucky (or unlucky), and consequently is expected to get closer to its expected line as the season progresses.

A team can overperform its Pythagorean winning percentage by winning a disproportionate number of close games. This claim can be confirmed by a brief data exploration. With the following code, we create a data frame (`results`) from the previously loaded 2011 game logs that contain the names of the teams and the runs scored. Two new columns are created: the variable `winner` contains the abbreviation of the winning team and a second variable `diff` contains the margin of victory.

```
results <- gl2011 |>
  select(
    VisitingTeam, HomeTeam,
    VisitorRunsScored, HomeRunsScore
  ) |>
  mutate(
    winner = ifelse(
      HomeRunsScore > VisitorRunsScored,
      HomeTeam, VisitingTeam
    ),
    diff = abs(VisitorRunsScored - HomeRunsScore)
  )
```

Suppose we focus on the games won by only one run. We create a data frame `one_run_wins` containing only the games decided by one run, and use the `n()` function to count the number of wins in such contests for each team.

```
one_run_wins <- results |>
  filter(diff == 1) |>
  group_by(winner) |>
  summarize(one_run_w = n())
```

Using the `my_teams` data frame previously created, we look at the relation between the Pythagorean residuals and the number of one-run victories. Note that the team abbreviation for the Angels needs to be changed because it is coded as `LAA` in the Lahman database and as `ANA` in the Retrosheet game logs.

```
teams2011 <- my_teams |>
  filter(yearID == 2011) |>
  mutate(
    teamID = if_else(teamID == "LAA", "ANA", as.character(teamID)
    )
  ) |>
  inner_join(one_run_wins, by = c("teamID" = "winner"))
```

The final bit of code produces the plot in Figure 4.3 which shows a positive relationship between the number of one-run games won and the Pythagorean residuals.

```
ggplot(data = teams2011, aes(x = one_run_w, y = residuals_pyt)) +
  geom_point() +
  geom_text_repel(aes(label = teamID)) +
  xlab("One run wins") + ylab("Pythagorean residuals")
```

Figure 4.3 shows that San Francisco had a large number of one-run victories and a large positive Pythagorean residual. In contrast, San Diego had few one-run victories and a negative residual.

Winning a disproportionate number of close games is sometimes attributed to plain luck. However, teams with certain attributes may be more likely to systematically win contests decided by a narrow margin. For example, teams with top quality closers will tend to preserve small leads, and will be able to overperform their Pythagorean expected winning percentage. To check this conjecture, we look at the data.

The `Pitching` table in the **Lahman** package contains individual seasonal pitching stats. We use the `filter()` function to select the pitcher-seasons where more than 50 games were finished by a pitcher with an ERA lower than 2.50. The data frame `top_closers` contains only the columns identifying the

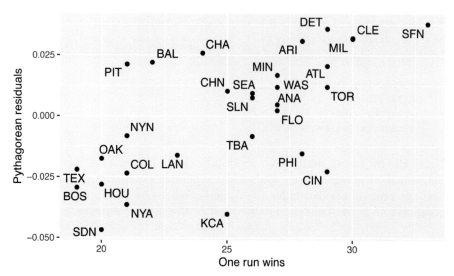

FIGURE 4.3
Scatterplot of number of one-run games won and Pythagorean residuals for major league teams in 2011.

pitcher, the season, and the team.

```
top_closers <- Pitching |>
  filter(GF > 50 & ERA < 2.5) |>
  select(playerID, yearID, teamID)
```

We merge the `top_closers` data frame with our `my_teams` dataset, creating a data frame that contains the teams featuring a top closer. We obtain summary statistics on the Pythagorean residuals using the `summary()` function.

```
my_teams |>
  inner_join(top_closers) |>
  pull(residuals_pyt) |>
  summary()
```

```
   Min. 1st Qu.  Median    Mean 3rd Qu.    Max.
-0.0487 -0.0109  0.0025  0.0054  0.0215  0.0812
```

The mean of the residuals is only slightly above zero (0.005), but when one multiplies it by the number of games in a season (162), one finds that teams with a top closer win, on average, 0.88 games more than would be predicted by the Pythagorean model.

4.5 How Many Runs for a Win?

Readers familiar with websites like http://www.insidethebook.com, http://www.hardballtimes.com, and http://www.baseballprospectus.com are surely familiar with the "ten-runs-equal-one-win" rule of thumb. Over the course of a season, a team scoring ten more runs is likely to have one more win in the standings. The number comes directly from the Pythagorean model with an exponent of two. Suppose a team scores an average of five runs per game, while allowing the same number of runs. In a 162-game season, the team would score (and allow) 810 runs. Inserting 810 in the Pythagorean formula one gets (as expected) a perfect .500 expected winning percentage with 81 wins. If one substitutes 810 with 820 for the number of runs scored in the formula, one obtains a .506 winning percentage that translates to 82 wins in 162 games. The same result is obtained for a team scoring 810 runs and allowing 800.

Ralph Caola derived the number of extra runs needed to get an extra win in a more rigorous way using calculus (Caola 2003). He starts from the equivalent representation of the Pythagorean formula.

$$W = G \cdot \frac{R^2}{R^2 + RA^2}$$

If one takes a *partial derivative* of the right side of the above equation with respect to R, holding RA constant, the result is the incremental number of wins per run scored. Taking the reciprocal of this result, one can derive the number of runs needed for an extra win.

R is capable of calculating partial derivatives, and thus we can retrace Ralph's steps in R by using the functions D() and expression() to take the partial derivative of $R^2/(R^2 + RA^2)$ with respect to R.

```
D(expression(G * R ^ 2 / (R ^ 2 + RA ^ 2)), "R")
```

```
G * (2 * R)/(R^2 + RA^2) - G * R^2 * (2 * R)/(R^2 + RA^2)^2
```

Unfortunately R does not do the simplifying. The reader has the choice of either doing the algebraic work herself or believing the final equation for incremental runs per win (IR/W) is the following[2]:

$$IR/W = \frac{\left(R^2 + RA^2\right)^2}{2 \cdot G \cdot R \cdot RA^2}$$

---

[2]The formula is the result of algebraic simplification and taking the reciprocal.

If R and RA are expressed in runs per game, we can remove G from the above formula.

Using this formula, one can compute the incremental runs needed per one win for various runs scored/runs allowed scenarios. As a first step, we create a function IR() to calculate the incremental runs, according to Caola's formula; this function takes runs scored per game and runs allowed per game as arguments.

```
IR <- function(RS = 5, RA = 5) {
  (RS ^ 2 + RA ^ 2)^2 / (2 * RS * RA ^ 2)
}
```

We use this function to create a table for various runs scored/runs allowed combinations. We perform this step by using the functions seq() and expand_grid(). The seq() function is used create a vector containing a regular sequence specifying, as arguments, the start value, the end value, and the increment value. Here seq() creates a vector of values from 3 to 6 in increments of 0.5. Then the expand_grid() function is used to obtain a data frame containing all the combinations of the elements of the supplied vectors. The following code displays the first and the final few lines of the new data frame ir_table.

```
ir_table <- expand_grid(
  RS = seq(3, 6, .5),
  RA = seq(3, 6, .5)
)
slice_head(ir_table, n = 4)
```

```
# A tibble: 4 x 2
     RS     RA
  <dbl> <dbl>
1     3     3
2     3   3.5
3     3     4
4     3   4.5
```

```
slice_tail(ir_table, n = 4)
```

```
# A tibble: 4 x 2
     RS     RA
  <dbl> <dbl>
1     6   4.5
2     6     5
3     6   5.5
4     6     6
```

Finally, we calculate the incremental runs for the various scenarios. The `pivot_wider()` function in the third line of the following code is used to show the results in a tabular form.

```
ir_table |>
  mutate(IRW = IR(RS, RA)) |>
  pivot_wider(
    names_from = RA, values_from = IRW, names_prefix = "RA="
  ) |>
  round(1)
```

```
# A tibble: 7 x 8
     RS `RA=3` `RA=3.5` `RA=4` `RA=4.5` `RA=5` `RA=5.5` `RA=6`
  <dbl>  <dbl>    <dbl>  <dbl>    <dbl>  <dbl>    <dbl>  <dbl>
1     3      6      6.1    6.5        7    7.7      8.5    9.4
2   3.5    7.2        7    7.1      7.5    7.9      8.5    9.2
3     4    8.7      8.1      8      8.1    8.4      8.8    9.4
4   4.5   10.6      9.6    9.1        9    9.1      9.4    9.8
5     5   12.8     11.3   10.5     10.1     10     10.1   10.3
6   5.5   15.6     13.4   12.2     11.4   11.1       11   11.1
7     6   18.8     15.8   14.1       13   12.4     12.1     12
```

Looking at the results, we notice that the rule of ten is appropriate in typical run scoring environments (4 to 5 runs per game). However, in very low scoring environments (the upper-left corner of the table), a lower number of runs is needed to gain an extra win; on the other hand, in high scoring environments (lower-right corner), one needs a larger number of runs for an added win.

4.6 Further Reading

Bill James first mentioned his Pythagorean model in (James 1980) which, like other early works by James, was self-published and is currently hard to find. Reference to the model is present in James (1982), the first edition published by Ballantine Books. Davenport and Woolner (1999) and Heipp (2003) revisited Bill James' model, deriving exponents that vary according to the total runs scored per game. Caola (2003) algebraically derived the relation between run scored and allowed and winning percentage. Star (2011) recounts the final moments of the 2011 regular season, when in the span of a few minutes the Rays and the Red Sox fates turned dramatically; the page also features a twelve-minute video chronicling the events of the wild September 28, 2011 night.

4.7   Exercises

1. Relationship Between Winning Percentage and Run Differential Across Decades

Section 4.3 used a simple linear model to predict a team's winning percentage based on its run differential. This model was fit using team data since the 2001 season.

- Refit this linear model using data from the seasons 1961–1970, the seasons 1971–1980, the seasons 1981–1990, and the seasons 1991–2000.
- Compare across the five decades the predicted winning percentage for a team with a run differential of 10 runs.

2. Pythagorean Residuals for Poor and Great Teams in the 19th Century

As baseball was evolving into its modern form, 19th century leagues often featured abysmal teams that did not even succeed in finishing their season, as well as some dominant clubs.

- Fit a Pythagorean formula model to the run differential, win-loss data for teams who played in the 19th century.
- By inspecting the residual plot of your fitted model from (a), did the great and poor teams in the 19th century do better or worse than one would expect on the basis of their run differentials?

3. Exploring the Manager Effect in Baseball

Retrosheet game logs report, for every game played, the managers of both teams.

- Select a period of your choice (encompassing at least ten years) and fit the Pythagorean formula model to the run-differential, win-loss data.

- On the basis of your fit in part (a) and the list of managers, compile a list of the managers who most overperformed their Pythagorean winning percentage and the managers who most underperformed it.

4. Pythagorean Relationship for Other Sports

Bill James' Pythagorean model has been used for predicting winning percentage in other sports. Since the pattern of scoring is very different among sports (compare for example points in basketball and goals in soccer), the model needs to be adapted to the scoring environment. Find the necessary data for a sport of your choice and compute the optimal exponent to the Pythagorean formula.

5

Value of Plays Using Run Expectancy

5.1 The Run Expectancy Matrix

An important concept in *sabermetrics* research is the run expectancy matrix. As each base (first, second, and third) can be either empty or occupied by a runner, there are $2 \times 2 \times 2 = 8$ possible arrangements of runners on the three bases. The number of outs can be 0, 1, or 2 (three possibilities), and so there are a total of $8 \times 3 = 24$ possible arrangements of runners and outs. For each combination of runners on base and outs, we are interested in computing the average number of runs scored in the remainder of the inning. When these average runs are arranged as a table classified by runners and outs, the display is often called the run expectancy matrix.

We use R to compute this matrix from play-by-play data for the 2016 season. This matrix is used to define the change in expected run value (often simply "run value") of a batter's plate appearance. We then explore the distribution of average run values for all batters in the 2016 season. The run values for José Altuve are used to help contextualize their meaning across players. We continue by exploring how players in different positions in the batting lineup perform with respect to this criterion. The notion of expected run value is helpful for understanding the relative benefit of different batting plays and we explore the value of a home run and a single. We conclude the chapter by using the run expectancy matrix and run values to understand the benefit of stealing a base and the cost of being caught stealing.

5.2 Runs Scored in the Remainder of the Inning

We begin by reading into R the play-by-play data we downloaded from Retrosheet for the 2016 season as `retro2016`. See Section A.1.3 for instructions for how to create the file `retro2016.rds`. (The retro2016 dataset is also available in the abdwr3edata package.)

```
retro2016 <- read_rds(here::here("data/retro2016.rds"))
```

At a given plate appearance, there is potential to score runs. Clearly, this potential is greater with runners on base, specifically runners in scoring position (second or third base), and when there are few outs. This potential for runs is estimated by computing the average number of runs scored in the remainder of the inning for each combination of runners on base and number of outs over some period of time. Certainly, the average runs scored depends on many variables such as home versus away, the current score, the pitching, and the defense. But this runs potential represents the opportunity to create runs in a typical situation during an inning and is a useful baseline against which to measure the contributions of players.

To compute the number of runs scored in the remainder of the inning, we need to know the total runs scored by both teams during the plate appearance and also the total runs scored by the teams at the end of the specific half-inning. The runs scored in the *remainder of the inning* (denoted by `runs_roi`) is the difference

$$runs_{roi} = runs_{\text{Total in Inning}} - runs_{\text{So far in Inning}}.$$

We create several new variables using the `mutate()` function: `runs_before` is equal to the sum of the visitor's score (`away_score_ct`) and the home team's score `home_score_ct` at each plate appearance, and `half_inning` uses the `paste()` function to combine the game id, the inning, and the team at bat, creating a unique identification for each half-inning of every game. Also, we create a new variable `runs_scored` that gives the number of runs scored for each play. (The variables `bat_dest_id`, `run1_dest_id`, `run2_dest_id`, and `run3_dest_id` give the destination bases for the batter and each runner, and runs are scored for each destination base that exceeds 3.)

```
retro2016 <- retro2016 |>
  mutate(
    runs_before = away_score_ct + home_score_ct,
    half_inning = paste(game_id, inn_ct, bat_home_id),
    runs_scored =
      (bat_dest_id > 3) + (run1_dest_id > 3) +
      (run2_dest_id > 3) + (run3_dest_id > 3)
  )
```

We wish to compute the maximum total score for each half-inning, combining home and visitor scores. We accomplish this by using the `summarize()` function, after grouping by `half_inning`. In the `summarize()` function, `outs_inning` is the number of outs for each half-inning, `runs_inning` is the total runs scored in each half-inning, `runs_start` is the score at the beginning of the half-inning, and `max_runs` is the maximum total score in a half-inning, which is the sum of the initial total runs and the runs scored. These summary data are stored in the new data frame `half_innings`.

```
half_innings <- retro2016 |>
  group_by(half_inning) |>
  summarize(
    outs_inning = sum(event_outs_ct),
    runs_inning = sum(runs_scored),
    runs_start = first(runs_before),
    max_runs = runs_inning + runs_start
  )
```

We use the `inner_join()` function to merge the data frames `data2016` and `half_innings`. Then the runs scored in the remainder of the inning (new variable `runs_roi`) can be computed by taking the difference of `max_runs` and runs.

```
retro2016 <- retro2016 |>
  inner_join(half_innings, by = "half_inning") |>
  mutate(runs_roi = max_runs - runs_before)
```

5.3 Creating the Matrix

Now that the runs scored in the remainder of the inning variable has been computed for each plate appearance, it is straightforward to compute the run expectancy matrix.

Currently, there are three variables `base1_run_id`, `base2_run_id`, and `base3_run_id` containing the player codes of the baserunners (if any) who are respectively on first, second, or third base. We create a new three-digit variable `bases` where each digit is either 1 or 0 if the corresponding base is respectively occupied or empty. The `state` variable adds the number of outs to the `bases` variable. One particular value of `state` would be "011 2", which indicates that there are currently runners on second and third base with two outs. A second state value "100 0" indicates there is a runner at first with no outs.

```
retro2016 <- retro2016 |>
  mutate(
    bases = paste0(
      if_else(base1_run_id == "", 0, 1),
      if_else(base2_run_id == "", 0, 1),
      if_else(base3_run_id == "", 0, 1)
```

```
    ),
    state = paste(bases, outs_ct)
  )
```

We want to only consider plays in our data frame where there is a change in the runners on base, number of outs, or runs scored. We create three new variables `is_runner1`, `is_runner2`, `is_runner3`, which indicate, respectively, if first base, second base, and third base are occupied after the play. (The function `as.numeric()` converts a logical variable to a numeric variable.) The variable `new_outs` is the number of outs after the play, `new_bases` indicates bases occupied, and `new_state` provides the runners on each base and the number of outs after the play.[1]

```
retro2016 <- retro2016 |>
  mutate(
    is_runner1 = as.numeric(
      run1_dest_id == 1 | bat_dest_id == 1
    ),
    is_runner2 = as.numeric(
      run1_dest_id == 2 | run2_dest_id == 2 |
        bat_dest_id == 2
    ),
    is_runner3 = as.numeric(
      run1_dest_id == 3 | run2_dest_id == 3 |
        run3_dest_id == 3 | bat_dest_id == 3
    ),
    new_outs = outs_ct + event_outs_ct,
    new_bases = paste0(is_runner1, is_runner2, is_runner3),
    new_state = paste(new_bases, new_outs)
  )
```

We use the `filter()` function to restrict our attention to plays where either there is a change between `state` and `new_state` (indicated by the not equal logical operator `!=` or there are runs scored on the play.

```
changes2016 <- retro2016 |>
  filter(state != new_state | runs_scored > 0)
```

Before the run expectancies are computed, one final adjustment is necessary. The play-by-play database includes scoring information for all half-innings during the 2016 season, including partial half-innings at the end of the game where the winning run is scored with less than three outs. In our computation

---

[1]The logic for computing the present and future state is encoded in the `retrosheet_add_states()` function in the **abdwr3edata** package. See the `run_expectancy_code()` function from the **baseballr** package for similar functionality.

of run expectancies, we want to work only with complete half-innings where three outs are recorded. We use the `filter()` function to extract the data from the half-innings in `changes2016` with exactly three outs—the new data frame is named `changes2016_complete`. (By removing the incomplete innings, we are introducing a small bias since these innings are not complete due to the scoring of at least one run.)

```
changes2016_complete <- changes2016 |>
  filter(outs_inning == 3)
```

We compute the expected number of runs scored in the remainder of the inning (the run expectancy) for each of the 24 bases/outs situations by use of the `summarize()` function, grouping by `bases` and `outs_ct` and employing the `mean()` function. We define store the resulting data frame as `erm_2016`.

```
erm_2016 <- changes2016_complete |>
  group_by(bases, outs_ct) |>
  summarize(mean_run_value = mean(runs_roi))
```

To display these run values as an 8 × 3 matrix, we use the `pivot_wider()` function.

```
erm_2016 |>
  pivot_wider(
    names_from = outs_ct,
    values_from = mean_run_value,
    names_prefix = "Outs="
  )
```

```
# A tibble: 8 x 4
# Groups:   bases [8]
  bases `Outs=0` `Outs=1` `Outs=2`
  <chr>    <dbl>    <dbl>    <dbl>
1 000      0.498    0.268    0.106
2 001      1.35     0.937    0.372
3 010      1.13     0.673    0.312
4 011      1.93     1.36     0.548
5 100      0.858    0.512    0.220
6 101      1.72     1.20     0.478
7 110      1.44     0.921    0.414
8 111      2.11     1.54     0.695
```

To see how the run expectancy values have changed over time, we input the 2002 season values as reported in Albert and Bennett (2003) in the matrix `erm_2002`. We display the 2016 and 2002 expectancies side-by-side for purposes

TABLE 5.1
Comparison of expected run values between 2016 (three columns labeled "NEW") and 2002 (three columns labeled "OLD").

	2016			2002		
bases	NEW=0	NEW=1	NEW=2	OLD=0	OLD=1	OLD=2
000	0.50	0.27	0.11	0.51	0.27	0.10
001	1.35	0.94	0.37	1.40	0.94	0.36
010	1.13	0.67	0.31	1.14	0.68	0.32
011	1.93	1.36	0.55	1.96	1.36	0.63
100	0.86	0.51	0.22	0.90	0.54	0.23
101	1.72	1.20	0.48	1.84	1.18	0.52
110	1.44	0.92	0.41	1.51	0.94	0.45
111	2.11	1.54	0.70	2.33	1.51	0.78

of comparison using `bind_cols()` in Table 5.1.

```
erm_2002 <- tibble(
  "OLD=0" = c(.51, 1.40, 1.14,  1.96, .90, 1.84, 1.51, 2.33),
  "OLD=1" = c(.27,  .94,  .68,  1.36, .54, 1.18,  .94, 1.51),
  "OLD=2" = c(.10,  .36,  .32,   .63, .23,  .52,  .45,  .78)
  )

out <- erm_2016 |>
  pivot_wider(
    names_from = outs_ct,
    values_from = mean_run_value,
    names_prefix = "NEW="
  ) |>
  bind_cols(erm_2002)
```

It is somewhat remarkable that these run expectancy values have not changed much over the recent history of baseball. This indicates there have been few changes in the average run scoring tendencies of MLB teams between 2002 and 2016.

5.4 Measuring Success of a Batting Play

When a player comes to bat with a particular runner and out situation, the run expectancy matrix tells us the average number of runs a team will score in the remainder of the half-inning. Based on the outcome of the plate appearance, the state (runners on base and outs) will change and there will be a updated run expectancy value. We estimate the value of the plate appearance, called the *run value*, by computing the difference in run expectancies of the old and new states plus the number of runs scored on the particular play.

$$\text{RUN VALUE} = \text{RUNS}_{\text{New state}} - \text{RUNS}_{\text{Old state}} + \text{RUNS}_{\text{Scored on Play}}$$

We compute the run values for all plays in the original data frame `retro2016` using the following R code. First, we use the `left_join()` function to match the expected run values for the beginning of each plate appearance. Note that we do this by matching the `bases` and `outs_ct` variables in `retro2016` to those in the run expectancy matrix `erm_2016`. This creates a new variable `mean_run_value`, which we promptly rename `rv_start`. Next, we do this again; this time matching the `new_bases` and `new_outs` variables to the run expectancy matrix to create the variable `rv_end`. It is important to use a `left_join()` (rather than an `inner_join()`) here, since the three out states are not present in `erm_2016`. The run expectancy of a situation with three outs is obviously zero, so we use the `replace_na()` function to set these run values to zero.

Thus, in the dataset `retro2016`, the variable `rv_start` is defined to be the run expectancy of the current state, and the variable `rv_end` is defined to be the run expectancy of the new state. The new variable `run_value` is set equal to the difference in `rv_end` and `rv_start` plus `runs_scored`.

```
retro2016 <- retro2016 |>
  left_join(erm_2016, join_by("bases", "outs_ct")) |>
  rename(rv_start = mean_run_value) |>
  left_join(
    erm_2016,
    join_by(new_bases == bases, new_outs == outs_ct)
  ) |>
  rename(rv_end = mean_run_value) |>
  replace_na(list(rv_end = 0)) |>
  mutate(run_value = rv_end - rv_start + runs_scored)
```

5.5 José Altuve

To better understand run values, let's focus on the plate appearances for the great hitter José Altuve for the 2016 season. To find Altuve's player id, we use the `People` data frame from the **Lahman** package and use the `filter()` function to extract the `retroID`. The `pull()` function extracts the vector `retroID` from the data frame.

```
library(Lahman)
altuve_id <- People |>
  filter(nameFirst == "Jose", nameLast == "Altuve") |>
  pull(retroID)
```

We then use the `filter()` function to isolate a data frame `altuve` of Altuve's plate appearances, where the batter id (variable `bat_id`) is equal to `altuve_id`. We wish to consider only the batting plays where Altuve was the hitter, so we also select the rows where the batting flag (variable `bat_event_fl`) is true.[2]

```
altuve <- retro2016 |>
  filter(
    bat_id == altuve_id,
    bat_event_fl == TRUE
  )
```

How did Altuve do in his first three plate appearances this season? To answer this, we display the first three rows of the data frame `altuve`, showing the original state, new state, and run value variables:

```
altuve |>
  select(state, new_state, run_value) |>
  slice_head(n = 3)
```

```
# A tibble: 3 x 3
  state new_state run_value
  <chr> <chr>         <dbl>
1 000 1 000 2       -0.162
2 000 1 100 1        0.244
3 000 1 000 2       -0.162
```

On his first plate appearance, there were no runners on base with one out. The outcome of this plate appearance was no runners on with two outs, indicating that Altuve got out, and the run value for this play was −0.162 runs. In his second plate appearance, the bases were again empty with one out. Here Altuve got on base, and the run value in the transition from "000 1" to "100 1" was 0.244 runs. In the third plate appearance, Altuve again got out in a bases empty, one-out situation, and the run value was −0.162 runs.

When one evaluates the run values for any player, there are two primary questions. First, we need to understand the player's opportunities for producing runs. What were the runner/outs situations for the player's plate appearances? Second, what did the batter do with these opportunities to score runs? The

---

[2]The variable `bat_event_fl` distinguishes batting events from non-batting events such as steals and wild pitches.

batter's success or lack of success on these opportunities can be measured in relation to these run values.

Let's focus on the runner states to understand Altuve's opportunities. Since a few of the counts of the runners/outs states over the 32 outcomes are close to zero, we focus on the runners on base variable `bases`. We apply the `n()` function within `summarize()` to tabulate the runners state for all of Altuve's plate appearances.

```
altuve |>
  group_by(bases) |>
  summarize(N = n())
```

```
# A tibble: 8 x 2
  bases      N
  <chr> <int>
1 000     417
2 001      24
3 010      60
4 011      18
5 100     128
6 101      22
7 110      40
8 111       8
```

We see that Altuve generally was batting with the bases empty (000) or with only a runner on first (100). Most of the time, Altuve was batting with no runners in scoring position.

How did Altuve perform with these opportunities? Using the `geom_jitter()` geometric object, we construct a jittered scatterplot that shows the run values for all plate appearances organized by the runners state (see Figure 5.1). Jittering the points in the horizontal direction is helpful in showing the density of run values. We also add a horizontal line at the value zero to the graph—points above the line (below the line) correspond to positive (negative) contributions.

```
ggplot(altuve, aes(bases, run_value)) +
  geom_jitter(width = 0.25, alpha = 0.5) +
  geom_hline(yintercept = 0, color = "red") +
  xlab("Runners on base")
```

When the bases were empty (000), the range of possible run values was relatively small. For this state, the large cluster of points at a negative run value corresponds to the many occurrences when Altuve got an out with the bases empty. The cluster of points at (000) at the value 1 corresponds to Altuve's home runs with the bases empty. (A home run with runners empty

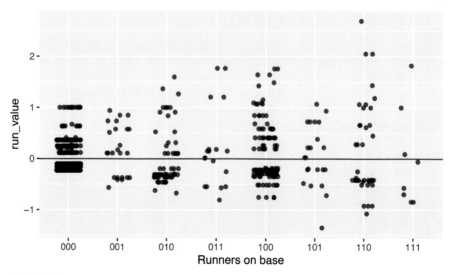

FIGURE 5.1

Stripchart of run values of José Altuve for all 2016 plate appearances as a function of the runners state. The points have been jittered since there are many plate appearances with identical run values.

will not change the bases/outs state and the value of this play is exactly one run.) For other situations, say the bases-loaded situation (111), there is much more variation in the run values. For one plate appearance, the state moved from 111 1 to 111 2, indicating that Altuve got out with the bases loaded with a run value of −0.84. In contrast, Altuve did hit a double with the bases loaded with one out, and the run value of this outcome was 1.82.

To understand Altuve's total run production for the 2016 season, we use the `summarize()` function together with the `sum()` and `n()` functions to compute the number of opportunities and sum of run values for each of the runners situations.

```
runs_altuve <- altuve |>
  group_by(bases) |>
  summarize(
    PA = n(),
    total_run_values = sum(run_value)
  )
runs_altuve
```

```
# A tibble: 8 x 3
  bases    PA total_run_values
  <chr> <int>            <dbl>
```

```
1 000      417           10.1
2 001       24            4.06
3 010       60            0.0695
4 011       18            3.43
5 100      128           10.2
6 101       22            1.34
7 110       40            5.62
8 111        8           -0.0968
```

We see, for example, that Altuve came to bat with the runners empty 417 times, and his total run value contribution to these 417 PAs was 10.10 runs. Altuve didn't appear to do particularly well with runners in scoring position. For example, there were 60 PAs where he came to bat with a runner on second base, and his net contribution in runs for this situation was 0.07 runs. Altuve's total runs contribution for the 2016 season can be computed by summing the last column of this data frame. This measure of batting performance is known as RE24, since it represents the change in run expectancy over the 24 base/out states (Appelman 2008).

```
    runs_altuve |>
      summarize(RE24 = sum(total_run_values))
```

```
# A tibble: 1 x 1
    RE24
   <dbl>
1   34.7
```

It is not surprising that Altuve has a positive total contribution in his PAs in 2016, but it is difficult to understand the size of 34.7 runs unless this value is compared with the contribution of other players. In the next section, we will see how Altuve compares to all hitters in the 2016 season.

5.6 Opportunity and Success for All Hitters

The run value estimates can be used to compare the batting effectiveness of players. We focus on batting plays, so we construct a new data frame retro2016_bat that is the subset of the main data frame retro2016 where the bat_event_fl variable is equal to TRUE:

```
    retro2016_bat <- retro2016 |>
      filter(bat_event_fl == TRUE)
```

It is difficult to compare the RE24 of two players at face value, since they have different opportunities to create runs for their teams. One player in the middle of the batting order may come to bat many times when there are runners in scoring position and good opportunities to create runs. Other players toward the bottom of the batting order may not get the same opportunities to bat with runners on base. One can measure a player's opportunity to create runs by the sum of the runs potential state (variable `rv_start`) over all of his plate appearances. We can summarize a player's batting performance in a season by the total number of plate appearances, the sum of the runs potentials, and the sum of the run values.

The R function `summarize()` is helpful in obtaining these summaries. We initially group the data frame `retro2016_bat` by the batter id variable `bat_id`, and for each batter, compute the total run value RE24, the total starting runs potential `runs_start`, and the number of plate appearances PA.

```
run_exp <- retro2016_bat |>
  group_by(bat_id) |>
  summarize(
    RE24 = sum(run_value),
    PA = length(run_value),
    runs_start = sum(rv_start)
  )
```

The data frame `run_exp` contains batting data for both pitchers and non-pitchers. It seems reasonable to restrict attention to non-pitchers, since pitchers and non-pitchers have very different batting abilities. Also we limit our focus on the players who are primarily starters on their teams. One can remove pitchers and non-starters by focusing on batters with at least 400 plate appearances. We create a new data frame `run_exp_400` by an application of the `filter()` function. We display the first few rows by use of the `slice_head()` function.

```
run_exp_400 <- run_exp |>
  filter(PA >= 400)
run_exp_400 |>
  slice_head(n = 6)
```

```
# A tibble: 6 x 4
  bat_id      RE24    PA runs_start
  <chr>      <dbl> <int>      <dbl>
1 abrej003    13.6   695       336.
2 alony001   -5.28   532       249.
3 altuj001    34.7   717       346.
4 andet001  -11.5    431       205.
5 andre001    17.7   568       257.
6 aokin001   -1.91   467       229.
```

Is there a relationship between batters' opportunities and their success in converting these opportunities to runs? To answer this question, we construct

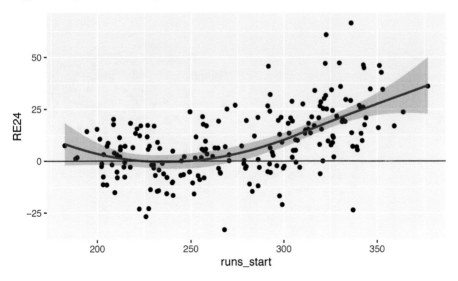

FIGURE 5.2
Scatterplot of total run value against the runs potential for all players in the 2016 season with at least 400 plate appearances. A smoothing curve is added to the scatterplot—this shows that players who had more run potential tend to have large run values.

a scatterplot of run opportunity (`runs_start`) against run value (`RE24`) for these hitters with at least 400 at bats (see Figure 5.2) using the `geom_point()` function. To help see the pattern in this scatterplot, we use the `geom_smooth()` function to add a *LOESS* smoother to the scatterplot. To interpret this graph, it is helpful to add a horizontal line (using the `geom_hline()` function) at 0—points above this line correspond to hitters who had a total positive run value contribution in the 2016 season.

```
plot1 <- ggplot(run_exp_400, aes(runs_start, RE24)) +
  geom_point() +
  geom_smooth() +
  geom_hline(yintercept = 0, color = "red")
plot1
```

From viewing Figure 5.2, we see that batters with larger values of `runs_start` tend to have larger runs contributions. But there is a wide spread in the run values for these players. In the group of players who have `runs_start` values between 300 and 350, four of these players actually have negative runs contributions and other players created over 60 runs in the 2016 season.

From the graph, we see that only a limited number of players created more than 40 runs for their teams. Who are these players? For labeling purposes, we extract the `nameLast` and `retroID` variables from the **Lahman People** data

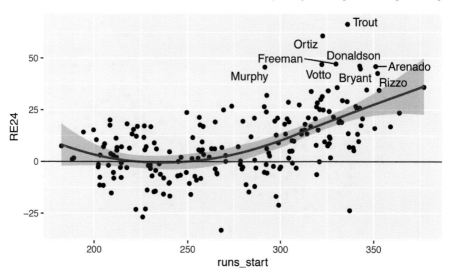

FIGURE 5.3
Scatterplot of RE24 against the runs potential for all players in the 2016 season
with unusual players identified.

frame and merge this information with the `run_exp_400` data frame. Using the
`geom_text_repel()` function, we add point labels to the previous scatterplot
for these outstanding hitters. This function from the **ggrepel** package plots the
labels so there is no overlap. (See Figure 5.3.)

```
run_exp_400 <- run_exp_400 |>
  inner_join(People, by = c("bat_id" = "retroID"))
library(ggrepel)
plot1 +
  geom_text_repel(
    data = filter(run_exp_400, RE24 >= 40),
    aes(label = nameLast)
  )
```

From Figure 5.3, we learn that the best hitters in terms of RE24 are Mike Trout
(66.43), David Ortiz (60.82), Freddie Freeman (47.19), Joey Votto (46.95), Josh
Donaldson (46.22), and Nolan Arenado (46.0).

5.7   Position in the Batting Lineup

Managers like to put their best hitters in the middle of the batting lineup.
Traditionally, a team's "best hitter" bats third and the cleanup hitter in the

fourth position is the best batter for advancing runners on base. What are the batting positions of the hitters in our sample? Specifically, are the best hitters using the run value criterion the ones who bat in the middle of the lineup?

A player may bat in several positions in the lineup during the season. We define a player's batting position as the position that he bats most frequently. We first merge the `retro2016` and `run_exp_400` data frames into the data frame `regulars`. Then by grouping `regulars` by the variables `bat_id` and `bat_lineup_id`, we find the frequency of each batting position for each player. Then by applications of the `arrange()` and `mutate()` functions, we define `position` to be the most frequent batting position. We add this new variable to the `run_exp_400` data frame.

```
regulars <- retro2016 |>
  inner_join(run_exp_400, by = "bat_id")

positions <- regulars |>
  group_by(bat_id, bat_lineup_id) |>
  summarize(N = n()) |>
  arrange(desc(N)) |>
  mutate(position = first(bat_lineup_id))

run_exp_400 <- run_exp_400 |>
  inner_join(positions, by = "bat_id")
```

In the following R code, the players' run opportunities are plotted against their RE24 values using `geom_text()` with `position` as the label variable. (See Figure 5.4.)

```
ggplot(run_exp_400, aes(runs_start, RE24, label = position)) +
  geom_text() +
  geom_hline(yintercept = 0, color = "red") +
  geom_point(
    data = filter(run_exp_400, bat_id == altuve_id),
    size = 4, shape = 16, color = crcblue
  )
```

From Figure 5.4, we better understand the relationship between batting position, run opportunities, and run values. The best hitters—the ones who create a large number of runs—generally bat third, fourth, and fifth in the batting order. The number of runs created by the leadoff (first) and second batters in the lineup are much smaller than the runs created by the best hitters in the middle (third and fourth positions) of the lineup. There are some surprises from this general pattern of batting positions. For example, there are some cleanup hitters (position 4) displayed who have mediocre values of runs created.

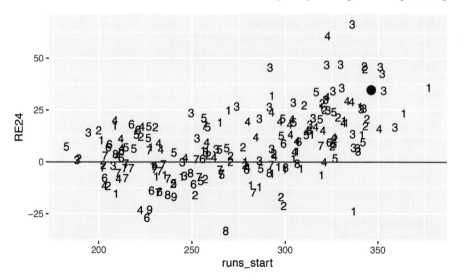

FIGURE 5.4
Scatterplot of total run value against the runs potential for all players in the
2016 season with at least 400 plate appearances. The points are labeled by the
position in the batting lineup and the large point corresponds to José Altuve.

How does José Altuve and his total run value of 34.7 compare among the
group of hitters with at least 400 plate appearances? We had saved Altuve's
batter id in the value `altuve_id`. Figure 5.4 uses another application of the
`geom_point()` function to display Altuve's (`runs_start, RE24`) value by a
large solid dot. In this particular season (2016), Altuve was one of the better
hitters in terms of creating runs for his team.

5.8   Run Values of Different Base Hits

There are many applications of run values in studying baseball. Here we look
at the value of a home run and a single from the perspective of creating runs.

One criticism of batting average is that it gives equal value to the four possible
base hits (single, double, triple, and home run). One way of distinguishing the
values of the base hits is to assign the number of bases reached: 1 for a single,
2 for a double, 3 for a triple, and 4 for a home run. Alternatively, slugging
percentage is the total number of bases divided by the number of at-bats. But
it is not clear that the values 1, 2, 3, and 4 represent a reasonable measure of
the value of the four possible base hits. We can get a better measure of the
importance of these base hits by the use of run values.

5.8.1 Value of a home run

Let's focus on the value of a home run from a runs perspective. We extract the home run plays from the data frame `retro2016` using the `event_cd` play event variable. An `event_cd` value of 23 corresponds to a home run. Using the `filter()` function with the `event_cd == 23` condition, we create a new data frame `home_runs` with the home run plays.

```
home_runs <- retro2016 |>
  filter(event_cd == 23)
```

What are the runners/outs states for the home runs hit during the 2016 season? We answer this question using the `table()` function.

```
home_runs |>
  select(state) |>
  table()
```

```
state
000 0 000 1 000 2 001 0 001 1 001 2 010 0 010 1 010 2 011 0
 1530   957   845    12    39    61    98   150   158    24
011 1 011 2 100 0 100 1 100 2 101 0 101 1 101 2 110 0 110 1
   37    39   319   357   340    28    74    63    82   131
110 2 111 0 111 1 111 2
  156    18    44    48
```

We compute the relative frequencies using the `prop.table()` function, and use the `round()` function to round the values to three decimal spaces.

```
home_runs |>
  select(state) |>
  table() |>
  prop.table() |>
  round(3)
```

```
state
000 0 000 1 000 2 001 0 001 1 001 2 010 0 010 1 010 2 011 0
0.273 0.171 0.151 0.002 0.007 0.011 0.017 0.027 0.028 0.004
011 1 011 2 100 0 100 1 100 2 101 0 101 1 101 2 110 0 110 1
0.007 0.007 0.057 0.064 0.061 0.005 0.013 0.011 0.015 0.023
110 2 111 0 111 1 111 2
0.028 0.003 0.008 0.009
```

We see from this table that the fraction of home runs hit with the bases empty is $0.273 + 0.171 + 0.151 = 0.595$. So over half of the home runs are hit with no runners on base.

Overall, what is the run value of a home run? We answer this question by computing the average run value of all the home runs in the data frame `home_runs`.

```
mean_hr <- home_runs |>
  summarize(mean_run_value = mean(run_value))
mean_hr
```

```
# A tibble: 1 x 1
  mean_run_value
           <dbl>
1           1.38
```

What are the run values of these home runs? We already observed in the analysis of Altuve's data that the run value of a home run with the bases empty is one. We construct a histogram of the run values for all home runs using the `geom_histogram()` function (see Figure 5.5).

```
ggplot(home_runs, aes(run_value)) +
  geom_histogram() +
  geom_vline(
    data = mean_hr, aes(xintercept = mean_run_value),
    color = "red", linewidth = 1.5
  ) +
  annotate(
    "text", 1.7, 2000,
    label = "Mean Run\nValue", color = "red"
  )
```

It is obvious from this graph that most home runs (the ones with the bases empty) have a run value of one. But there is a cluster of home runs with values between 1.5 and 2.0, and there is a small group of home runs with run values exceeding three.

Which runner/out situations lead to the most valuable home runs? Using the `arrange()` function, we display the row of the data frame corresponding to the largest run value.

```
home_runs |>
  arrange(desc(run_value)) |>
  select(state, new_state, run_value) |>
  slice_head(n = 1)
```

```
# A tibble: 1 x 3
  state new_state run_value
  <chr> <chr>         <dbl>
```

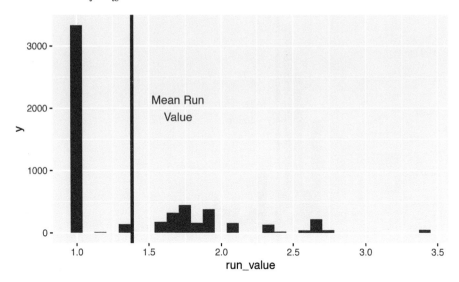

FIGURE 5.5
Histogram of the run values of the home runs hit during the 2016 season. The vertical line shows the location of the mean run value of a home run.

```
1 111 2 000 2         3.41
```

As one might expect, the most valuable home run occurs when there are bases loaded with two outs. The run value of this home run is 3.41.

Using the `geom_vline()` function, we draw a vertical line on the graph showing the mean run value and a label to this line (see Figure 5.5). This average run value is pretty small in relation to the value of a two-out grand slam, but this value partially reflects the fact that most home runs are hit with the bases empty.

5.8.2 Value of a single

Run values can also be used to evaluate the benefit of a single. Unlike a home run, the run value of a single will depend both on the initial state (runners and outs) and on the final state. The final state of a home run will always have the bases empty; in contrast, the final state of a single will depend on the movement of any runners on base.

We use the `filter()` function to select the plays where `event_cd` equals 20 (corresponding to a single); the new data frame is called `singles`. We construct a histogram of the run values for all of the singles in the 2016 season in Figure 5.6. As in the case of the home run, it is straightforward to compute the mean run value of a single. We display this mean value on the histogram in Figure 5.6.

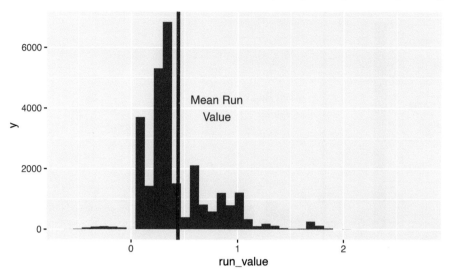

FIGURE 5.6
Histogram of the run values of the singles hit during the 2016 season. The
vertical line shows the location of the mean run value of a single.

```
singles <- retro2016 |>
  filter(event_cd == 20)

mean_singles <- singles |>
  summarize(mean_run_value = mean(run_value))

ggplot(singles, aes(run_value)) +
  geom_histogram(bins = 40) +
  geom_vline(
    data = mean_singles, color = "red",
    aes(xintercept = mean_run_value), linewidth = 1.5
  ) +
  annotate(
    "text", 0.8, 4000,
    label = "Mean Run\nValue", color = "red"
  )
```

Looking at the histogram of run values of the single, there are three large
spikes between 0 and 0.5. These large spikes can be explained by constructing
a frequency table of the beginning state.

```
singles |>
  select(state) |>
```

```
table()
```

```
state
000 0 000 1 000 2 001 0 001 1 001 2 010 0 010 1 010 2 011 0
 6920  4767  3763    71   323   354   516   745   864    90
011 1 011 2 100 0 100 1 100 2 101 0 101 1 101 2 110 0 110 1
  224   208  1665  1974  1765   159   321   368   364   697
110 2 111 0 111 1 111 2
  729   115   280   257
```

We see that most of the singles occur with the bases empty, and the three tall spikes in the histogram, as one moves from left to right in Figure 5.6, correspond to singles with no runners on and two outs, one out, and no outs. The small cluster of run values in the interval 0.5 to 2.0 correspond to singles hit with runners on base.

What is the most valuable single from the run value perspective? We use the `arrange()` function to find the beginning and end states for the single that resulted in the largest run value.

```
singles |>
  arrange(desc(run_value)) |>
  select(state, new_state, run_value) |>
  slice_head(n = 1)
```

```
# A tibble: 1 x 3
  state new_state run_value
  <chr> <chr>         <dbl>
1 111 2 001 2         2.68
```

In this particular play, the hitter came to bat with the bases loaded and two outs, and the final state was a runner on third with two outs. How could this have happened with a single? The data frame does contain a brief description of the play. But from the data frame we identify the play happening during the bottom of the 8th inning of a game between the Orioles and Yankees on June 5, 2016. We check with Baseball-Reference to find the following play description:

> Single to CF (Ground Ball thru SS-2B); Trumbo Scores; Davis Scores; Pena Scores/unER/Adv on E8 (throw); to 3B/Adv on throw

So evidently, the center fielder made an error on the fielding of the single that allowed all three runners to score and the batter to reach third base.

At the other extreme, by another use of the `arrange()` function, we identify two plays that achieved the smallest run value.

```
  singles |>
    arrange(run_value) |>
    select(state, new_state, run_value) |>
    slice(1)
```

```
# A tibble: 1 x 3
  state new_state run_value
  <chr> <chr>        <dbl>
1 010 0 100 1         -0.621
```

How could the run value of a single be negative six tenths of a run? With further investigation, we find that in each case, there was a runner on second who was hit by the ball in play and was called out.

In this case, we see that the mean value of a single is approximately equal to the run value when a single is hit with the bases empty with no outs. It is interesting that the run value of a single can be large (in the 1 to 2 range). These large run values reflect the fact that the benefit of the single depends on the advancement of the runners.

5.9   Value of Base Stealing

The run expectancy matrix is also useful in understanding the benefits of stealing bases. When a runner attempts to steal a base, there are two likely outcomes—either the runner will be successful in stealing the base or the runner will be caught stealing. Overall, is there a net benefit to attempting to steal a base?

The variable `event_cd` gives the code of the play and codes of 4 and 6 correspond respectively to a stolen base (SB) or caught stealing (CS). Using the `filter()` function, we create a new data frame `stealing` that consists of only the plays where a stolen base is attempted.

```
  stealing <- retro2016 |>
    filter(event_cd %in% c(4, 6))
```

By use of the `summarize()` and `n()` functions, we find the frequencies of the SB and CS outcomes.

```
  stealing |>
    group_by(event_cd) |>
    summarize(N = n()) |>
```

```
      mutate(pct = N / sum(N))
```

```
# A tibble: 2 x 3
  event_cd      N   pct
     <int> <int> <dbl>
1        4  2213 0.756
2        6   713 0.244
```

Among all stolen base attempts, the proportion of stolen bases is equal to 2213 / (2213 + 713) = 0.756.

What are common runners/outs situations for attempting a stolen base? We answer this by constructing a frequency table for the state variable.

```
    stealing |>
      group_by(state) |>
      summarize(N = n())
```

```
# A tibble: 16 x 2
    state     N
    <chr> <int>
 1 001 1      1
 2 001 2      1
 3 010 0     37
 4 010 1    124
 5 010 2    102
 6 011 1      1
 7 100 0    559
 8 100 1    708
 9 100 2    870
10 101 0     37
11 101 1     99
12 101 2    219
13 110 0     30
14 110 1     84
15 110 2     53
16 111 1      1
```

We see that stolen base attempts typically happen with a runner only on first (state 100). But there are a wide variety of situations where runners attempt to steal.

Every stolen base attempt has a corresponding run value that is stored in the variable run_value. This run value reflects the success of the attempt (either SB or CS) and the situation (runners and outs) where this attempt occurs. Using the geom_histogram() function, we construct a histogram of all of the

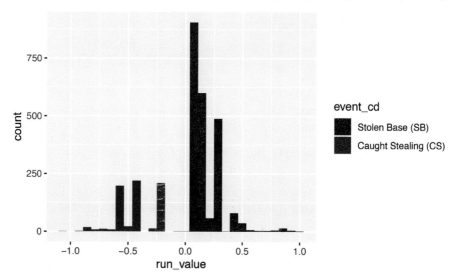

FIGURE 5.7
Histogram of the run values of all steal attempts during the 2016 season.

runs created for all the stolen base attempts in Figure 5.7. The color of the
bar indicates the success or failure of the attempt.

```
ggplot(stealing, aes(run_value, fill = factor(event_cd))) +
  geom_histogram() +
  scale_fill_manual(
    name = "event_cd",
    values = crc_fc,
    labels = c("Stolen Base (SB)", "Caught Stealing (CS)")
  )
```

Generally, all of the successful SBs have positive run value, although most of
the values fall in the interval from 0 to 0.3. In contrast, the unsuccessful CSs
(as expected) have negative run values. In further exploration, one can show
that the three spikes for negative run values correspond to CS when there is
only a runner on first with 0, 1, and 2 outs.

Let's focus on the benefits of stolen base attempts in a particular situation.
We create a new data frame that gives the attempted stealing data when there
is a runner on first base with one out (state "100 1").

```
stealing_1001 <- stealing |>
  filter(state == "100 1")
```

By tabulating the `event_cd` variable, we see the runner successfully stole 498
times out of 498 + 210 attempts for a success rate of 70.3%.

```
stealing_1001 |>
  group_by(event_cd) |>
  summarize(N = n()) |>
  mutate(pct = N / sum(N))
```

```
# A tibble: 2 x 3
  event_cd     N   pct
     <int> <int> <dbl>
1        4   498 0.703
2        6   210 0.297
```

Another way to look at the outcome is to look at the frequencies of the new_state variable.

```
stealing_1001 |>
  group_by(new_state) |>
  summarize(N = n()) |>
  mutate(pct = N / sum(N))
```

```
# A tibble: 4 x 3
  new_state     N    pct
    <chr>     <int>  <dbl>
1 000 1         1 0.00141
2 000 2       211 0.298
3 001 1        39 0.0551
4 010 1       457 0.645
```

This provides more information than simply recording a stolen base. On 457 occurrences, the runner successfully advanced to second base. On an additional 39 occurrences, the runner advanced to third. Perhaps this extra base was due to a bad throw from the catcher or a misplay by the infielder. More can be learned about the details of these plays by further examination of the other variables.

We are most interested in the value of attempting stolen bases in this situation— we address this by computing the mean run value of all of the attempts with a runner on first with one out.

```
stealing_1001 |>
  summarize(Mean = mean(run_value))
```

```
# A tibble: 1 x 1
     Mean
    <dbl>
1 0.00723
```

Stolen base attempts are worthwhile, although the value overall is about 0.007 runs per attempt. Of course, the actual benefit of the attempt depends on the success or failure and on the situation (runners and outs) where the stolen base is attempted.

5.10   Further Reading and Software

Lindsey (1963) was the first researcher to analyze play-by-play data in the manner described in this chapter. Using data collected by his father for the 1959–60 season, Lindsey obtained the run expectancy matrix that gives the average number of runs in the remainder of the inning for each of the runners/outs situations. Chapters 7 and 9 of Albert and Bennett (2003) illustrate the use of the run expectancy matrix to measure the value of different base hits and to assess the benefits of stealing and sacrifice hits. Tango, Lichtman, and Dolphin (2007), in their Toolshed chapter, describe the run expectancy table as one of the fundamental tools used throughout their book. Also, run expectancy plays a major role in the essays in Keri and Baseball Prospectus (2007).

Benjamin S. Baumer, Jensen, and Matthews (2015) use these run expectancies in the computation of WAR (wins above replacement) measures for players. The website http://www.fangraphs.com/library/misc/war/ introduces WAR, a useful way of summarizing a player's total contribution to his team.

5.11   Exercises

1. Run Values of Hits

In Section 5.8, we found the average run value of a home run and a single.

(a) Use similar R code as described in Section 5.8 for the 2016 season data to find the mean run values for a double, and for a triple.

(b) Albert and Bennett (2003) use a regression approach to obtain the weights 0.46, 0.80, 1.02, and 1.40 for a single, double, triple, and home run, respectively. Compare the results from Section 5.8 and part (a) with the weights of Albert and Bennett.

2. Value of Different Ways of Reaching First Base

There are three different ways for a runner to get on base: a single, walk (BB), or hit-by-pitch (HBP). But these three outcomes have different run values due to the different advancement of the runners on base. Use run values based on data from the 2016 season to compare the benefit of a walk, a hit-by-pitch, and a single when there is a single runner on first base.

3. Comparing Two Players with Similar OBPs

Adam Eaton (Retrosheet batter id `eatoa002`) and Starling Marte (Retrosheet batter id `marts002`) both had 0.362 on-base percentages during the 2016 season. By exploring the run values of these two payers, investigate which player was really more valuable to his team. Can you explain the difference in run values in terms of traditional batting statistics such as AVG, SLG, or OBP?

4. Create Probability of Scoring a Run Matrix

In Section 5.3, we illustrate the construction of the run expectancy matrix from 2016 season data. Suppose instead that one was interested in computing the proportion of times when at least one run was scored for each of the 24 possible bases/outs situations. Use R to construct this probability of scoring matrix.

5. Runner Advancement with a Single

Suppose one is interested in studying how runners move with a single.

a. Using the `filter()` function, select the plays when a single was hit. (The value of `event_cd` for a single is 20.) Call the new data frame `singles`.

b. Use the `group_by()` and `summarize()` functions with the data frame `singles` to construct a table of frequencies of the variables `state` (the beginning runners/outs state) and `new_state` (the final runners/outs state).

c. Suppose there is a single runner on first base. Using the table from part (b), explore where runners move with a single. Is it more likely for the lead runner to move to second, or to third base?

d. Suppose instead there are runners on first and second. Explore where runners move with a single. Estimate the probability a run is scored on the play.

6. Hitting Evaluation of Players by Run Values

Choose several players who were good hitters in the 2016 season. For each player, find the run values and the runners on base for all plate appearances. As in Figure 5.1, construct a graph of the run values against the runners on base. Was this particular batter successful when there were runners in scoring position?

6

Balls and Strikes Effects

6.1 Introduction

In this chapter we explore the effect of the ball/strike count on the behavior of players and umpires and on the final outcome of a plate appearance. We use Retrosheet data from the 2016 season to estimate how the ball/strike count affects the run expectancy. We also use Statcast data to explore how one pitcher modifies his pitch selection, how one batter alters his swing zone, and how umpires judge pitches based on the count. Along the way, we introduce functions for string manipulation that are useful for managing the pitch sequences from the Retrosheet play-by-play files. Level plots and contour plots, created with the use of the **ggplot2** package, will be used for the explorations of batters' swing tendencies and umpires' strike zones.

6.2 Hitter's Counts and Pitcher's Counts

When watching a broadcast of a baseball game, one often hears an announcer's concern for a pitcher who is repeatedly "falling behind" in the count, or his/her anticipation for a particular pitch because it's a "hitter's count" and the batter has a chance to do some damage. We will see if there is actual evidence that the so-called hitter's count really leads to more favorable outcomes for batters, while "getting ahead" in the count (a pitcher's count) is beneficial for pitchers.

6.2.1 An example for a single pitcher

The Baseball-Reference[1] website provides various splits for every player—in particular, it gives splits by ball/strike counts for all seasons since 1988. We find Mike Mussina's split statistics by entering the player's profile page (typing "Mussina" on the search box brings one there), clicking on the "Splits" tab in the "Standard Pitching" table, and clicking on "Career" (or whatever

---

[1]https://baseball-reference.com

season we are interested in) on the pop-up menu that appears. One finds the "Count Balls/Strikes" table scrolling down on the splits page. Alternatively, the table can be reached by a direct link: in this case the career splits by count for Mike Mussina are currently available at http://www.baseball-reference.com/players/split.fcgi?id=mussimi01&year=Career&t=p#count.

The first series of lines (from "First Pitch" to "Full Count") shows the statistics for events happening in that particular count. Thus, for example, a batting average (BA) of .338 on 1-0 counts indicates batters hit safely 34% of the time when putting the ball in play *on a 1-0 count* against Mussina. We are more interested in the second group of rows, those beginning with the word "After". In fact, in these cases the statistics are relative to every plate appearance that *goes through* that count. Thus a .337 on-base percentage (OBP) after 1-0 means that, whenever Mike Mussina started a batter with a ball, the batter successfully got on base 34% of the time, no matter how many pitches it took to end the plate appearance.

The last column on every table in the splits page is tOPS+. It's an index for comparing the player's OPS in that particular situation (the sum of on-base percentage and slugging percentage[2]) to his overall rself_index("OPS")'. A value over 100 is interpreted as a higher OPS in the situation compared to his OPS on all counts; conversely, values below 100 indicate a OPS value that is lower than his overall OPS.

Figure 6.1 uses a heat map to display Mussina's tOPS+ through the various counts. If one focuses on a particular number of strikes (moving up vertically), a higher number of balls in the count makes the outcome more likely to be favorable to the hitter (darker shades). Conversely, if one fixes the number of balls (moving right horizontally), the balance moves toward the pitcher (lighter shades) as one increases the number of strikes.

Figure 6.1 emphasizes the importance from a pitcher's perspective of beginning the duel with a strike. When Mussina fell behind 1-0 in his career, batters performed 18% better than usual in the plate appearance; conversely, after a first pitch strike, they were limited to 72% of their potential performance.

How is the heat map display of Figure 6.1 created in R? First we prepare a data frame mussina with all the possible balls/strikes counts, using the expand_grid() function as previously illustrated in Section 4.5. We then add a new variable value with the tOPS+ values taken from the Baseball-Reference website.

```
mussina <- expand_grid(balls = 0:3, strikes = 0:2) |>
  mutate(
    value = c(
```

---

[2]OPS is widely used as a measure of offensive production because—while being very easy to calculate—it correlates very well with runs scored at the team level.

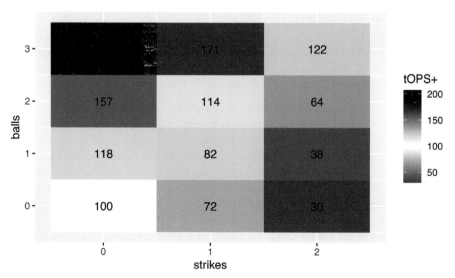

FIGURE 6.1
Heat map of tOPS+ for Mike Mussina through each ball/strike count. Data
from Baseball-Reference website.

```
        100, 72, 30, 118, 82, 38,
        157, 114, 64, 207, 171, 122
      )
    )
  mussina
```

```
# A tibble: 12 x 3
    balls strikes value
    <int>   <int> <dbl>
  1     0       0   100
  2     0       1    72
  3     0       2    30
  4     1       0   118
  5     1       1    82
  6     1       2    38
  7     2       0   157
  8     2       1   114
  9     2       2    64
 10     3       0   207
 11     3       1   171
 12     3       2   122
```

We create a **ggplot2** called `count_plot` by mapping strikes to the x aesthetic,
balls to the y aesthetic, and `fill` color to the `value`. The tiles that we draw

are colored based on this value. We use the `scale_fill_gradient2()` function to set a diverging color palette for the value of tOPS+. Since 100 is a neutral value, we set that to the midpoint and assign that value to the color white. For reasons that will become clear later on, we round the labels displayed in each tile (even though they are integers!).

```
count_plot <- mussina |>
  ggplot(aes(x = strikes, y = balls, fill = value)) +
  geom_tile() +
  geom_text(aes(label = round(value, 3))) +
  scale_fill_gradient2(
    "tOPS+", low = "grey10", high = crcblue,
    mid = "white", midpoint = 100
  )
count_plot
```

6.2.2 Pitch sequences from Retrosheet

From viewing Figure 6.1, we obtain an initial view of hitter's counts (darker shades) and pitcher's counts (lighter shades) on the basis of offensive production. However this figure is based on data for a single pitcher—does a similar pattern emerge when using league-wide data?

Retrosheet provides pitch sequences beginning with the 1988 season. Sequences are stored in strings such as `FBSX`. Each character encodes the description of a pitch. In this example, the pitch sequence is a foul ball (`F`), followed by a ball (`B`), a swinging strike (`S`), and the ball put into play (`X`). Table 6.1 provides the description for every code used in Retrosheet pitch sequences.[3]

6.2.2.1 Functions for string manipulation

Sequence strings from Retrosheet often require some initial processing before they can be suitably analyzed. In this section we provide a quick tutorial on some R functions for the manipulation of strings. Readers not interested in string manipulation functions may skip to Section 6.2.3.

The function `str_length()` returns the number of characters in a string. This function is helpful for obtaining the number of pitches delivered in a Retrosheet pitch sequence. For example, the number of pitches in the sequence `BBSBFFFX` is given by

```
str_length("BBSBFFFX")
```

```
[1] 8
```

---

[3]Source: https://www.retrosheet.org/eventfile.htm.

TABLE 6.1

Pitch codes used by Retrosheet.

Symbol	Description
+	following pickoff throw by the catcher
*	indicates the following pitch was blocked by the catcher
.	marker for play not involving the batter
1	pickoff throw to first
2	pickoff throw to second
3	pickoff throw to third
>	indicates a runner going on the pitch
B	ball
C	called strike
F	foul
H	hit batter
I	intentional ball
K	strike (unknown type)
L	foul bunt
M	missed bunt attempt
N	no pitch (on balks and interference calls)
O	foul tip on bunt
P	pitchout
Q	swinging on pitchout
R	foul ball on pitchout
S	swinging strike
T	foul tip
U	unknown or missed pitch
V	called ball because pitcher went to his mouth
X	ball put into play by batter
Y	ball put into play on pitchout

However, as indicated in Table 6.1, there are some characters in the Retrosheet strings denoting actions that are not pitches, such as pickoff attempts.

The functions `str_which()` and `str_detect()` are used to find patterns within elements of character vectors. The function `str_which()` returns the indices of the elements for which a match is found, and the function `str_detect()` returns a logical vector, indicating for each element of the vector whether a match is found. For both functions, the first argument is the vector of strings where matches are sought and the second argument is the string pattern to search for. For example, we apply the two functions to the vector of pitch sequences `sequences` to search for pickoff attempts to first base denoted by the code `1`.

```
sequences <- c("BBX", "C11BBC1S", "1X")
str_which(sequences, "1")
```

[1] 2 3

```
str_detect(sequences, "1")
```

[1] FALSE   TRUE   TRUE

The function `str_which()` tells us that "1" is contained in the second (2) and third (3) components of the character vector `sequences`, and `str_detect()` outputs this same information by means of a logical vector.

The pattern to search for does not have to be a single character. In fact, it can be any *regular expression*. For example we may want to look for consecutive pickoff attempts to first, which is the pattern "11". The output below shows that "11" is contained in the second component of `sequences`.

```
str_detect(sequences, "11")
```

[1] FALSE   TRUE FALSE

The function `str_replace_all()` allows for the substitution of the pattern found with a replacement. The replacement can be an empty string, in which case the pattern is simply removed[ˆstr_remove]. In the code below, we use the function `str_remove_all()` to removes the pickoff attempts to first from the pitch sequences.

```
str_remove_all(sequences, "1")
```

[1] "BBX"   "CBBCS" "X"

6.2.2.2   Finding plate appearances going through a given count

Since we are interested only in pitch counts, we should remove the characters not corresponding to actual pitches from the pitch sequences. Regular expressions are the computing tool needed for this particular task. While it's beyond the scope of this book to fully explain how regular expressions work, we will instead show a few examples on how to use them.[4]

We begin by loading the `retro2016.rds` file containing Retrosheet's play-by-play for the 2016 season using the `read_rds()` function. Please see Section A.1.3

---

[4]The website https://www.regular-expressions.info is a comprehensive online resource on regular expressions, featuring examples, tutorials, references for syntax, and a list of related books.

for instructions on how to create the file `retro2016.rds`.

```
retro2016 <- read_rds(here::here("data/retro2016.rds"))
```

We use the `str_remove_all()` function to create the variable `pseq` of pitch sequences after removing the symbols from the Retrosheet pitch sequence variable `pitch_seq_tx` that don't correspond to actual pitches.

```
pbp2016 <- retro2016 |>
  mutate(pseq = str_remove_all(pitch_seq_tx, "[.>123N+*]"))
```

In a regular expression, the square brackets indicate the collection of characters to search. The above code removes pickoff attempts at any base (1, 2, 3) either by the pitcher or the catcher (+), balks and interference calls (N), plays not involving the batter (.), indicators of runners going on the pitch (>), and of catchers blocking the pitch (*).[5]

We need another special character to identify the plate appearances that go through a 1-0 count. In a regular expression, the ^ character means the pattern has to be matched at the beginning of the string. Looking at Table 6.1 there are four different ways a ball can be coded (B, I, P, V). A plate appearance that goes though a 1-0 count must therefore begin with one of these characters. The following code creates the desired variable c10.

```
pbp2016 <- pbp2016 |>
  mutate(c10 = str_detect(pseq, "^[BIPV]"))
```

Similarly, plate appearances going through a 0-1 count must start with a strike. Thus, we create a new variable c01.

```
pbp2016 <- pbp2016 |>
  mutate(c01 = str_detect(pseq, "^[CFKLMOQRST]"))
```

Let's check our work by checking the values of `pitch_seq_tx`, c10, and c01 for the first ten lines of the data frame.

```
pbp2016 |>
  select(pitch_seq_tx, c10, c01) |>
  slice_head(n = 10)
```

```
# A tibble: 10 x 3
  pitch_seq_tx c10    c01
```

---

[5]Note that applying the `str_length()` function to the newly created variable `pseq` gives the number of pitches delivered in each at-bat.

```
     <chr>          <lgl> <lgl>
 1 BX              TRUE  FALSE
 2 X               FALSE FALSE
 3 SFS             FALSE TRUE
 4 BCX             TRUE  FALSE
 5 BSS*B1S         TRUE  FALSE
 6 BBX             TRUE  FALSE
 7 BCX             TRUE  FALSE
 8 CX              FALSE TRUE
 9 BCCS            TRUE  FALSE
10 SBFX            FALSE TRUE
```

Writing regular expressions for every pitch count is a tedious task and we will refer the reader to Section A.3 for the full code.

6.2.3 Expected run value by count

In order to compute the expected runs values by count, we need to augment pbp2016 via three steps:

1. We need to compute the base-out state, based on the configuration of the baserunners and the number of outs, both for the beginning and end of the play. This process is identical to the one explicated in Section 5.3. To avoid duplication of code, in this chapter we use the function `retrosheet_add_states()`, which we put in the **abdwr3edata** package for this purpose.
2. Using the state computed above, we need to join the expected run matrix that we created in Section 5.3 for the 2016 season. This provides us with the expected run values for both the beginning and end states of each play.
3. We need to compute variables analogous to `c01` and `c10` as described above, but for each of the 12 possible counts. We have placed that code in the `retrosheet_add_counts()` function in the **abdwr3edata** package. See Section A.3 for more details on how to compute these new variables.

For step 1, we simply call the `retrosheet_add_states()` function.

```
library(abdwr3edata)
pbp2016 <- pbp2016 |>
  retrosheet_add_states()
```

For step 2, we load the 2016 expected run matrix and repeatedly join it onto our play-by-play data, similar to what we did in Section 5.4.

```
erm_2016 <- read_rds(here::here("data/erm2016.rds"))

pbp2016 <- pbp2016 |>
```

```
left_join(
  erm_2016, join_by(bases, outs_ct)
) |>
rename(rv_start = mean_run_value) |>
left_join(
  erm_2016, join_by(new_bases == bases, new_outs == outs_ct)
) |>
rename(rv_end = mean_run_value) |>
replace_na(list(rv_end = 0)) |>
mutate(run_value = rv_end - rv_start + runs_scored)
```

For step 3, we call `retrosheet_add_counts()`.

```
pbp2016 <- pbp2016 |>
  retrosheet_add_counts()
```

Now we have a beginning and ending run value associated with each play, and we know whether the at-bat moved through each of the 12 possible counts.

```
pbp2016 |>
  select(
    game_id, event_id, run_value, c00, c10, c20,
    c11, c01, c30, c21, c31, c02, c12, c22, c32
  ) |>
  slice_head(n = 5)
```

```
# A tibble: 5 x 15
  game_id event_id run_value c00   c10   c20   c11   c01   c30
  <chr>      <int>     <dbl> <lgl> <lgl> <lgl> <lgl> <lgl> <lgl>
1 ANA201~        1     0.635 TRUE  TRUE  FALSE FALSE FALSE FALSE
2 ANA201~        2    -0.196 TRUE  FALSE FALSE FALSE FALSE FALSE
3 ANA201~        3    -0.565 TRUE  FALSE FALSE FALSE TRUE  FALSE
4 ANA201~        4     0.848 TRUE  TRUE  FALSE TRUE  FALSE FALSE
5 ANA201~        5    -0.220 TRUE  TRUE  FALSE TRUE  FALSE FALSE
# i 6 more variables: c21 <lgl>, c31 <lgl>, c02 <lgl>,
#   c12 <lgl>, c22 <lgl>, c32 <lgl>
```

For example, the at-bat in the fourth line of the data frame (a two-out RBI single) started with a 1-0 count (value TRUE in column c10), then moved to the count 1-1, and finally generated a change in expected runs of 0.848. The pbp2016 data frame has all the necessary information to calculate the run values of the various balls/strikes counts, in the same way the value of a home run and of a single were calculated in Section 5.8.

As an illustration, we can measure the importance of getting ahead on the first pitch. We calculate the mean change in expected run value for at-bats starting with a ball and for the at-bats starting with a strike.

```
pbp2016 |>
  filter(c10 == 1 | c01 == 1) |>
  group_by(c10, c01) |>
  summarize(
    num_ab = n(),
    mean_run_value = mean(run_value)
  )
```

```
# A tibble: 2 x 4
# Groups:   c10 [2]
  c10   c01   num_ab mean_run_value
  <lgl> <lgl>  <int>          <dbl>
1 FALSE TRUE   94106        -0.0394
2 TRUE  FALSE  76165         0.0371
```

The conclusion is that the difference between a first pitch strike and a first pitch ball, as estimated with data from the 2016 season, is over 0.07 expected runs.

We can calculate the run value for each possible ball/strike count. First, we use the select() function and the starts_with() operator to grab only those columns that start with the letter c. In this case, this matches the columns c00, c01, etc. that we defined previously. Additionally, we grab the run_value column.

```
pbp_counts <- pbp2016 |>
  select(starts_with("c"), run_value)
```

Now, we want to apply the group_by()-summarize() idiom that we used previously to calculate the mean run value across *all* of the 12 possible counts. One way to do this would be to write a function that will perform that operation for a given variable name, and then iterate over the 12 variable names, and indeed, that is the approach taken in the first edition of this book. Here, we employ an alternative strategy that is more in keeping with the **tidyverse** philosophy and involves much less code. However, it may be conceptually less intuitive.

pbp_counts has $n = 190713$ rows and $p = 13$ columns. The variable named run_value contains a measurement of runs, and the other 12 columns contain logical indicators as to whether the plate appearance passed through a particular count. Thus, we really have three different kinds of information in this data frame: a count, whether the plate appearance passed through that count, and the run value. To *tidy* these data (Wickham 2014), we need to create a long data frame with $12n$ rows and those three columns. We do this using the pivot_longer() function. We provide a name to the names_to argument, which becomes the name of the new variable that records the count. Similarly, the `values_to` argument takes a name for the new variable that records the data that was in the column that was gathered. We don't want to gather the `run_value` column, since that records a different kind of information.

```
pbp_counts_tidy <- pbp_counts |>
  pivot_longer(
    cols = -run_value,
    names_to = "count",
    values_to = "passed_thru"
  )
pbp_counts_tidy |>
  sample_n(6)
```

```
# A tibble: 6 x 3
  run_value count passed_thru
      <dbl> <chr> <lgl>
1    -0.220 c00   TRUE
2     0.206 c12   FALSE
3     0.360 c32   FALSE
4    -0.106 c01   TRUE
5    -0.230 c22   TRUE
6    -0.106 c01   FALSE
```

Note that every plate appearance appears $p = 12$ times in `pbp_counts_tidy`: one row for each count. To compute the average change in expected runs (i.e., the mean run value), we have to `filter()` for only those plate appearances that actually passed through that count. Then we simply apply our `group_by()`-`summarize()` operation as before. Thus, in the `mean()` operation, the data is limited to only those plate appearances that have gone through each particular ball-strike count.

```
run_value_by_count <- pbp_counts_tidy |>
  filter(passed_thru) |>
  group_by(count) |>
  summarize(
    num_ab = n(),
    value = mean(run_value)
  )
```

Finally, we can then update our `count_plot` to use these new data instead of the old `mussina` data. To do this, we have to re-compute the balls and strikes based on the count variable. We can do this by picking out the values of balls and strikes using the `str_sub()` function. We use the `scale_fill_gradient2()` function again to reset our diverging palette to colors more appropriate for these data (i.e., a midpoint at 0).

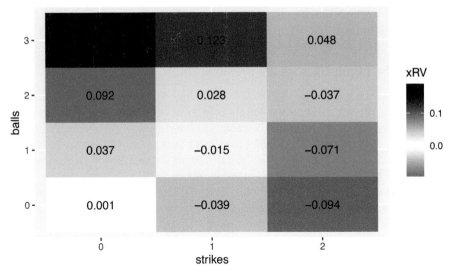

FIGURE 6.2
Average change in expected runs for plate appearances passing through each
ball/strike count. Values estimated on data from the 2016 season.

```
run_value_by_count <- run_value_by_count |>
  mutate(
    balls = str_sub(count, 2, 2),
    strikes = str_sub(count, 3, 3)
  )

count_plot %+%
  run_value_by_count +
  scale_fill_gradient2(
    "xRV", low = "grey10", high = crcblue,
    mid = "white"
  )
```

By glancing at the values and color shades in Figure 6.2, one can construct
reasonable definitions for the terms "hitter's count" and "pitcher's count". Note
that since all plate appearances pass through the 0-0 count, the average change
in expected run value for this count is approximately 0. Ball/strike counts can
be roughly divided in the following four categories[6]:

- Pitcher's counts: 0-2, 1-2, 2-2, 0-1;
- Neutral counts: 0-0, 1-1;

---

[6]The proposed categorization, based on observation of Figure 6.2, reflects the one proposed
by analyst Tom Tango (see https://www.insidethebook.com/ee/index.php/site/comments/p
late_counts/).

- Modest hitter's counts: 3-2, 2-1, 1-0;
- Hitter's counts: 3-0, 3-1, 2-0.

6.2.4 The importance of the previous count

In the previous section we calculated run values for any ball/strike count. In performing this calculation we simply looked at whether a plate appearance went through a particular count, without considering how it got there. In other words, we considered, for example, all the at-bats going through a 2-2 count as having the same run expectancy, no matter if the pitcher started ahead 0-2 or fell behind 2-0. The implicit assumption in these calculations is that the previous counts have no influence on the outcome on a particular count[7]. However, a pitcher who got ahead 0-2 is likely to "waste some pitches". That is, he would likely throw a few balls out of the strike zone with the sole intent of inducing the batter (who cannot afford another strike) to swing at them and possibly miss or make poor contact. On the other hand, with a plate appearance starting with two balls, the batter has the luxury of not swinging at strikes in undesirable locations and waiting for the pitcher to deliver a pitch of his liking.

Given the above discussion, it would seem that the run expectancy on a 2-2 count would be higher if the plate appearance started with two balls than if the pitcher started with a 0-2 count. Let's investigate if there is numerical evidence to actually support this conjecture.

We begin by taking the subset of plays from the 2016 season that went through a 2-2 count and calculating their mean change in expected run value.

```
count22 <- pbp2016 |>
  filter(c22 == 1)
count22 |>
  summarize(
    num_ab = n(),
    mean_run_value = mean(run_value)
  )
```

```
# A tibble: 1 x 2
  num_ab mean_run_value
   <int>          <dbl>
1  44254        -0.0368
```

Using the `case_when()` function, we create a new variable `after2` denoting the ball/strike count after two pitches and calculate the mean run value for each of the three possible levels of `after2`.

---

[7]This is analogous to the *memoryless property* referred to in Section 9.2.1.

```
count22 |>
  mutate(
    after2 = case_when(
      c20 == 1 ~ "2-0",
      c02 == 1 ~ "0-2",
      c11 == 1 ~ "1-1",
      TRUE ~ "other")
  ) |>
  group_by(after2) |>
  summarize(
    num_ab = n(),
    mean_run_value = mean(run_value)
  )
```

```
# A tibble: 3 x 3
  after2 num_ab mean_run_value
  <chr>   <int>          <dbl>
1 0-2      9837        -0.0311
2 1-1     28438        -0.0376
3 2-0      5979        -0.0422
```

The above results appear to imply that plate appearances going through a 2-2 count after having started with two strikes are more favorable to the batter than those beginning with two balls.

This result should be considered in light of multiple types of potential *selection bias*. Many plate appearances starting with two strikes end without ever reaching the 2-2 count, in most cases with an unfavorable outcome for the batter.[8] The plate appearances that survive a 0-2 count reaching 2-2 are hardly a random sample of all the plate appearances. Hard-to-strike-out batters are likely over-represented in such a sample, as well as pitchers with good fastball command (to get ahead 0-2), but weak secondary pitches (to finish off batters).

Similarly, comparing the paths leading to 1-1 counts yields results in line with common sense.

```
count11 <- pbp2016 |>
  filter(c11 == 1)
count11 |>
  mutate(
    after2 = ifelse(c10 == 1, "1-0", "0-1")
  ) |>
  group_by(after2) |>
  summarize(
```

---

[8] In 2016, 85% of plate appearances beginning with two strikes and not reaching the 2-2 count ended with the batter making an out.

```
    num_ab = n(),
    mean_run_value = mean(run_value)
)
```

```
# A tibble: 2 x 3
  after2 num_ab mean_run_value
  <chr>  <int>          <dbl>
1 0-1    38759        -0.0122
2 1-0    36467        -0.0176
```

The numbers above suggest that after reaching a 1-1 count, the batter is expected to perform slightly worse if the first pitch was a ball than if it was a strike.

6.3 Behaviors by Count

In this section we explore how the roles of three individuals in the pitcher-batter duel are affected by the ball/strike count. How does a batter alter his swing when ahead or behind in the count? How does a pitcher vary his mix of pitches according to the count? Does an umpire (consciously or unconsciously), shrink or expand his strike zone depending on the pitch count?

We provide an R data file (a file with extension .Rdata) containing all the datasets used in this section. Once the file is loaded into R, the data frames `cabrera`, `sanchez`, `umpires`, and `verlander` are visible by use of the `ls()` function.

```
load(here::here("data/balls_strikes_count.RData"))
```

These datasets contain pitch-by-pitch data, including the location of pitches as recorded by Sportvision's PITCHf/x system. The `cabrera` data frame contains four years of batting data for 2012 American League Triple Crown winner Miguel Cabrera. The data frame `umpires` has information about every pitch thrown in 2012 where the home plate umpire had to judge whether it crossed the strike zone. The `verlander` data frame has four years of pitching data for 2016 Cy Young Award and MVP recipient Justin Verlander.

6.3.1 Swinging tendencies by count

We saw in Section 6.2 that batters perform worse when falling behind in the count. For example, when there are two strikes in the count, the batter may be forced to swing at pitches he would normally let pass by to avoid being

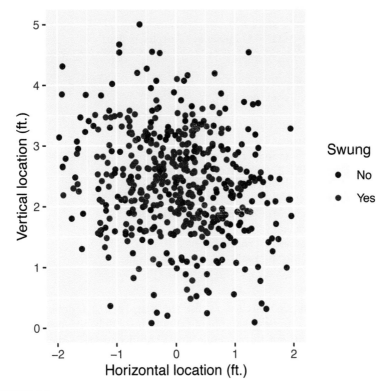

FIGURE 6.3
Scatterplot of Miguel Cabrera's swinging tendency by location. Sample of 500 pitches. View from the catcher's perspective.

called out on strikes. Using PITCHf/x data, we explore how a very good batter like Miguel Cabrera alters his swinging tendencies according to the ball/strike count.

6.3.1.1 Propensity to swing by location

In this section we focus on the relationships between the variables `balls` and `strikes` indicating the count on the batter, the variables `px` and `pz` identifying the pitch location as it crosses the front of the plate, and the `swung` binary variable, denoting whether or not the batter attempted a swing on the pitch.

We show a scatterplot of Miguel Cabrera's swinging tendency in Figure 6.3.

```
k_zone_plot <- cabrera |>
  sample_n(500) |>
  ggplot(aes(x = px, y = pz)) +
  geom_rect(
```

```
    xmin = -0.947, xmax = 0.947, ymin = 1.5,
    ymax = 3.6, fill = "lightgray", alpha = 0.01
  ) +
  coord_equal() +
  scale_x_continuous(
    "Horizontal location (ft.)",
    limits = c(-2, 2)
  ) +
  scale_y_continuous(
    "Vertical location (ft.)",
    limits = c(0, 5)
  )
k_zone_plot +
  geom_point(aes(color = factor(swung))) +
  scale_color_manual(
    "Swung", values = crc_fc,
    labels = c("No", "Yes")
  )
```

Rather than plot all 6265 pitches, we use the `sample_n()` function to simplify matters by taking a random sample of 500 pitches. This reduces overlapping in the scatterplot without introducing bias. From Figure 6.3, one can see that Cabrera is less likely to swing at pitches delivered farther away from the strike zone (the black box). However, it is difficult to determine Cabrera's preferred pitch location from this figure.

A contour plot is an effective alternative method to visualize batters' swinging preferences. This type of plot is used to visualize three-dimensional data in two dimensions. Widely used in cartography and meteorology, the contour plot usually features spatial coordinates as the first two variables, while the third variable (which can be, for example, elevation in cartography or barometric pressure in meteorology) is plotted as a contour line, also called an isopleth. The contour line is a curve joining points sharing equal values of the third variable.

As a first step in producing a contour plot, we fit a smooth polynomial surface to the response variable `swung` as a function of the horizontal and vertical locations `px` and `pz` using the `loess()` function. The output of this fit is stored in the object `miggy_loess`.

```
miggy_loess <- loess(
  swung ~ px + pz,
  data = cabrera,
  control = loess.control(surface = "direct")
)
```

After the surface has been fit, we are interested in predicting the likelihood of a swing by Cabrera at various pitch locations. Using the `expand_grid()` function, we build a data frame consisting of combinations of horizontal locations from -2 (two feet to the left of the middle of home plate) to $+2$ (two feet to the right of the middle of the plate) and vertical locations from the ground (value of zero) to six feet of height, using subintervals of 0.1 feet. Using the `predict()` function, we obtain the likelihood of Miguel's swinging at every location in the data frame. Note that because `predict()` returns a `matrix`, we use the `as.numeric()` function to convert the fitted values into a numeric vector.

```
pred_area <- expand_grid(
  px = seq(-2, 2, by = 0.1),
  pz = seq(0, 6, by = 0.1)
)
pred_fits <- miggy_loess |>
  predict(newdata = pred_area) |>
  as.numeric()
pred_area_fit <- pred_area |>
  mutate(fit = pred_fits)
```

To spot check, we examine in the data frame `pred_area_fit` the likelihood that Cabrera will swing for three different hand-picked locations—a pitch down the middle and two and a half feet from the ground ("down *Broadway*", a ball that hits the ground in the middle of the plate ("ball in the dirt"), and another one delivered at mid-height (2.5 feet from the ground) but way outside (two feet from the middle of the plate). In each case, we use the `filter()` function to take a subset of the prediction data frame `pred_area_fit` with specific values of the horizontal and vertical locations px and pz.

```
pred_area_fit |>
  filter(px == 0 & pz == 2.5)   # down Broadway
```

```
# A tibble: 1 x 3
    px    pz   fit
 <dbl> <dbl> <dbl>
1    0   2.5 0.844
```

```
pred_area_fit |>
  filter(px == 0 & pz == 0)      # ball in the dirt
```

```
# A tibble: 1 x 3
    px    pz   fit
 <dbl> <dbl> <dbl>
1    0     0 0.154
```

```
    pred_area_fit |>
      filter(px == 2 & pz == 2.5)   # way outside


# A tibble: 1 x 3
     px    pz    fit
  <dbl> <dbl>  <dbl>
1     2   2.5 0.0783
```

The results are quite consistent with what one would expect: the pitch right in the heart of the strike zone induces Cabrera to swing more than 80 percent of the time, while the ball in the dirt and the ball outside generate a swing at 15 percent and 8 percent rates, respectively.

We construct a contour plot of the likelihood of the swing as a function of the horizontal and vertical locations of the pitch using the `geom_contour2()` function in the **metR** package. For logical consistency, we `filter()` for only those contours corresponding to swing probabilities between 0 and 1. Figure 6.4 shows the resulting contour plot.

```
    cabrera_plot <- k_zone_plot %+%
      filter(pred_area_fit, fit >= 0, fit <= 1) +
      metR::geom_contour2(
        aes(
          z = fit, color = after_stat(level),
          label = after_stat(level)
        ),
        binwidth = 0.2, skip = 0
      ) +
      scale_color_gradient(low = "white", high = crcblue)

    cabrera_plot
```

As expected, the likelihood of a swing decreases the further the ball is delivered from the middle of the strike zone. The plot also shows that Cabrera has a tendency to swing at pitches on the inside part of the plate.

6.3.1.2 Effect of the ball/strike count

Figure 6.4 reports Cabrera's swinging tendency over all pitch counts. Can we visualize how Cabrera varies his approach according to the ball/strike count? Specifically, does Cabrera become more selective when he is ahead and can afford to wait for a pitch of his liking and, conversely, does he "expand his zone" when there are two strikes and he cannot allow another called strike go by? We described the process of calculating the swing propensity by location in Section 6.3.1.1. Here, we generalize that procedure and iterate it over all counts.

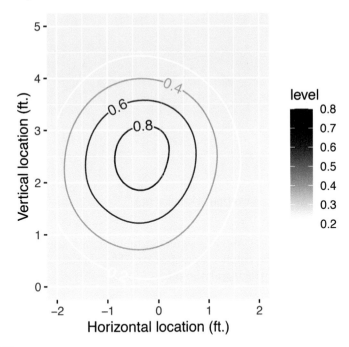

FIGURE 6.4
Contour plot of Miguel Cabrera's swinging tendency by location, where the view is from the catcher's perspective. The contour lines are labeled by the probability of swinging at the pitch.

In this case, we restrict our interest to 0-0, 0-2, and 2-0 counts. The vector `counts` contains these values. Next, we split the `cabrera` data frame into a `list` with three elements: one data frame for each of the chosen counts. We accomplish this by `filter()`-ing for those `counts` and using the `group_split()` function to do the splitting. Note that the resulting `count_dfs` object is a `list` of three `data.frames`.

```
counts <- c("0-0", "0-2", "2-0")
count_dfs <- cabrera |>
  mutate(count = paste(balls, strikes, sep = "-")) |>
  filter(count %in% counts) |>
  group_split(count)
```

Next, we use the `map()` function repeatedly to iterate our analysis over the elements of `count_dfs`. First, we compute the LOESS fits for a given set of data specific to the count using `loess()`. Second, we use `predict()` to compute three sets of predictions—one for each of the three counts. Third, we convert the numeric matrices returned by `predict()` to numeric vectors

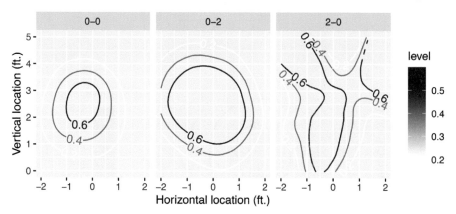

FIGURE 6.5

Contour plots of Miguel Cabrera's swinging tendency by selected ball/strike counts, viewed from the catcher's perspective. The contour lines are labeled by the probability of swinging at the pitch.

called `fit`, and append them to the `pred_area` data frame. Fourth, we use `set_names()` to link the counts with their corresponding data frames. Finally, we stitch all three data frames together using `list_rbind()`, and add variables for the count, number of balls, and number of strikes.

```
count_fits <- count_dfs |>
  map(~loess(
    swung ~ px + pz, data = .x,
    control = loess.control(surface = "direct")
  )) |>
  map(predict, newdata = pred_area) |>
  map(~tibble(pred_area, fit = as.numeric(.x))) |>
  set_names(nm = counts) |>
  list_rbind(names_to = "count") |>
  mutate(
    balls = str_sub(count, 1, 1),
    strikes = str_sub(count, 3, 3)
  )
```

This process performs the same tasks that we did before: it fits a LOESS model to the pitch location data, then uses that model to generate swing probability predictions across the entire area and returns a `tibble` with the associated ball and strike count.

We can then use a `facet_wrap()` to show the contour plots on separate panels to compare Cabrera's swinging tendencies by pitch count (see Figure 6.5). To improve legibility, we only show the 20%, 40%, and 60% contours.

```
cabrera_plot %+%
  filter(count_fits, fit > 0.1, fit < 0.7) +
  facet_wrap(vars(count))
```

As expected, Cabrera expands his swing zone when behind 0-2 (his 40% contour line on 0-2 counts has an area comparable to his 20% contour line on 0-0 counts). The third panel in Figure 6.5 is comprised of a relatively low sample size and probably does not tell us much.

6.3.2 Pitch selection by count

We now move to the other side of the pitcher/batter duel in our investigation of the effect of the count. Pitchers generally possess arsenals of two to five different pitch types. All pitchers have a fastball at their disposal, which is generally a pitch that is easy to throw to a desired location. So-called secondary pitches, such as curve balls or sliders, while often effective (especially when hitters are not expecting them), are harder to control and rarely used by pitchers behind in the count. In this section we look at one pitcher (arguably one of the best in MLB at the time) and explore how he chooses from his pitch repertoire according to the ball/strike count.

The `verlander` data frame, consisting of over 15 thousand observations, consists of pitch data for Justin Verlander for four seasons. Using the `group_by()` and `summarize()` commands, we obtain a frequency table of the types of pitches Verlander threw from 2009–2012. In this case, we compute the pitch type proportions in addition to their frequencies.

```
verlander |>
  group_by(pitch_type) |>
  summarize(N = n()) |>
  mutate(pct = N / nrow(verlander)) |>
  arrange(desc(pct))
```

```
# A tibble: 5 x 3
  pitch_type     N    pct
  <fct>       <int>  <dbl>
1 FF           6756  0.441
2 CU           2716  0.177
3 CH           2550  0.167
4 FT           2021  0.132
5 SL           1264  0.0826
```

As is the case with most major league pitchers, Verlander throws his fastball most frequently. He uses two variations of a fastball: a four-seamer (FF) and a two-seamer (FT). He complements his fastballs with a curve ball (CU), a change-up (CH), and a slider (SL).

We see in the table that 44% of Verlander's pitches during this four-season period were four-seam fastballs.

Before moving to exploring pitch selection by ball/strike count, we compute a frequency table to explore the pitch selection by batter handedness. The `pivot_wider()` function helps us display the results in a wide rather than long format.

```
verlander |>
  group_by(batter_hand, pitch_type) |>
  summarize(N = n()) |>
  pivot_wider(names_from = batter_hand, values_from = N) |>
  mutate(L_pct = L / sum(L), R_pct = R / sum(R))
```

```
# A tibble: 5 x 5
  pitch_type     L     R L_pct  R_pct
  <fct>      <int> <int> <dbl>  <dbl>
1 CH          2024   526 0.228 0.0817
2 CU          1529  1187 0.172 0.184
3 FF          3832  2924 0.432 0.454
4 FT          1303   718 0.147 0.111
5 SL           178  1086 0.0201 0.169
```

Note that Verlander's pitch selection is quite different depending on the handedness of the opposing batter. In particular, the right-handed Verlander uses his changeup nearly a quarter of the time against left-handed hitters, but only eight percent of the time against right-handed hitters. Conversely the slider is nearly absent from his repertoire when he faces lefties, while he uses it close to one out of six times against righties.

Batter-hand differences in pitch selection are common among major league pitchers and they exist because the effectiveness of a given pitch depends on the handedness of the pitcher and the batter. The slider and change-up comparison is a typical example, a slider is very effective against batters of the same handedness while a change-up can be successful when facing opposite-handed batters.

We can also explore Verlander's pitch selection by pitch count as well as batter handedness. In the following code, the `filter()` function is used to select Verlander's pitches delivered to right-handed batters. The rest of the code constructs a table of frequencies by count and pitch type. The `across()` function helps to divide each numeric variable by the total number of pitches.

```
verlander |>
  filter(batter_hand == "R") |>
  group_by(balls, strikes, pitch_type) |>
```

```
    summarize(N = n()) |>
    pivot_wider(
      names_from = pitch_type,
      values_from = N,
      values_fill = 0
    ) |>
    mutate(
      num_pitches = CH + CU + FF + FT + SL,
      across(where(is.numeric), ~.x / num_pitches)
    ) |>
    select(-num_pitches)
```

```
# A tibble: 12 x 7
# Groups:   balls, strikes [12]
   balls strikes     CH     CU     FF      FT     SL
   <int>   <int>  <dbl>  <dbl>  <dbl>   <dbl>  <dbl>
 1     0       0 0.0692  0.115  0.526  0.156   0.134
 2     0       1 0.0624  0.242  0.405  0.101   0.189
 3     0       2 0.158   0.282  0.274  0.0629  0.223
 4     1       0 0.0493  0.107  0.523  0.112   0.209
 5     1       1 0.0805  0.238  0.398  0.0960  0.187
 6     1       2 0.143   0.327  0.278  0.0617  0.191
 7     2       0 0.0174  0.0174 0.703  0.145   0.116
 8     2       1 0.0623  0.0796 0.512  0.142   0.204
 9     2       2 0.102   0.294  0.357  0.0869  0.161
10     3       0 0.0833  0      0.812  0.104   0
11     3       1 0.0196  0      0.784  0.118   0.0784
12     3       2 0.0429  0.0429 0.693  0.116   0.106
```

The effect of the ball/strike count on the choice of pitches is apparent when comparing pitcher's counts and hitter's counts. When behind 2-0, Verlander uses his four-seamer seven times out of ten; the percentage goes up to 78% when trailing 3-1 and 81% on 3-0 counts. Conversely, when he has the chance to strike the batter out, the use of the four-seamer diminishes. In fact he throws it less than 30 percent of the time both on 0-2 and 1-2 counts. On a full count, Verlander's propensity to throw his fastball is similar to those of hitters' counts—this is consistent with the numbers in Figure 6.2 that indicate the 3-2 count being slightly favorable to the hitter. One can explore Verlander's choices by count when facing a left-handed hitter by simply changing R to L in the code above.

6.3.3 Umpires' behavior by count

Hardball Times author John Walsh wrote a 2010 article titled *The Compassionate Umpire* in which he showed that home plate umpires tend to modify

TABLE 6.2

A twenty row sample of the `umpires` dataset.

Umpire	batter_hand	balls	strikes	px	pz	called_strike
Bill Miller	R	2	1	0.84	1.44	0
Jeff Nelson	L	0	0	−1.58	2.64	0
James Hoye	R	0	0	0.65	1.36	0
Tim Tschida	L	0	1	0.83	1.34	0
Fieldin Culbreth	L	1	2	−1.22	2.37	0
Tim McClelland	L	1	1	−0.67	2.03	1
Laz Diaz	R	0	1	0.43	3.35	1
Paul Nauert	L	0	0	−1.42	4.04	0
James Hoye	L	0	0	−0.27	2.89	1
Kerwin Danley	R	1	1	−0.84	4.86	0
Mike Everitt	L	0	0	−0.03	1.62	1
Dan Bellino	R	0	0	0.78	0.62	0
Mike Everitt	R	0	0	−0.93	1.22	0
Tom Hallion	R	1	0	0.39	2.38	1
Jerry Meals	L	0	2	−0.84	1.41	0
Tim Welke	R	0	0	−1.72	1.34	0
James Hoye	L	1	0	−1.97	2.99	0
Brian O'Nora	R	0	0	−0.57	3.69	0
Jordan Baker	R	0	1	0.00	3.91	0
Gerry Davis	R	0	1	−1.02	4.91	0

their ball/strike calling behavior by slightly favoring the player who is behind in the count (Walsh 2010). In other words, umpires tend to enlarge their strike zone in hitter's counts and to shrink it when pitchers are ahead. In this section we visually explore Walsh's finding by plotting contour lines for three different ball/strike counts.

The `umpires` data frame is similar to those of `verlander` and `cabrera`. A sample of its contents—obtained using the `sample_n()` function—is shown in Table 6.2.

```
sample_n(umpires, 20)
```

The data consist of every pitch of the 2012 season for which the home plate umpire had to judge whether it crossed the strike zone. Additional columns not present in either the `verlander` or the `cabrera` data frames identify the name of the umpire (variable `umpire`) and whether the pitch was called for a strike (variable `called_strike`).

We proceed similarly to the analysis of Section 6.3.1.2, using the `loess()` function to estimate the umpires' likelihood of calling a strike, based on the location of the pitch. Here we limit the analysis to plate appearances featuring right-handed batters, as it has been shown that umpires tend to call pitches slightly differently depending on the handedness of the batter.

```
umpires_rhb <- umpires |>
  filter(
    batter_hand == "R",
    balls == 0 & strikes == 0 |
      balls == 3 & strikes == 0 |
      balls == 0 & strikes == 2
  )
```

By slightly modifying the code above, the reader can easily repeat the process for other counts. In this section we compare the 0-0 count to the most extreme batter and pitcher counts, 3-0 and 0-2 counts, respectively.

To do this, we can re-purpose our `map()` pipeline from above, incorporating the `pred_area` data frame. Note that the response variable in the LOESS model here is `called_strike`. In addition, the `loess()` smoother is applied on a subset of 3000 randomly selected pitches, to reduce computation time.

```
ump_counts <- umpires_rhb |>
  mutate(count = paste(balls, strikes, sep = "-")) |>
  group_by(count)

counts <- ump_counts |>
  group_keys() |>
  pull(count)

ump_count_fits <- ump_counts |>
  group_split() |>
  map(sample_n, 3000) |>
  map(~loess(
    called_strike ~ px + pz, data = .x,
    control = loess.control(surface = "direct"))
  ) |>
  map(predict, newdata = pred_area) |>
  map(~tibble(pred_area, fit = as.numeric(.x))) |>
  set_names(nm = counts) |>
  list_rbind(names_to = "count") |>
  mutate(
    balls = str_sub(count, 1, 1),
    strikes = str_sub(count, 3, 3)
  )
```

Figure 6.6 shows that the umpire's strike zone shrinks considerably in a 0-2 pitch count, and slightly expanded in a 3-0 count. To isolate the 0.5 contour lines, we

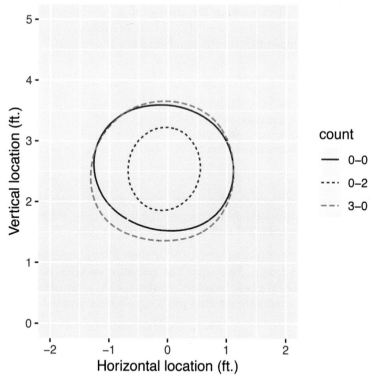

FIGURE 6.6
Umpires' 50/50 strike calling zone in different balls/strikes counts viewed from the catcher's perspective.

`filter()` the fitted values for those near 0.5, and then use `geom_contour()` to set the width of the bins to be small.

```
k_zone_plot %+% filter(ump_count_fits, fit < 0.6 & fit > 0.4) +
  geom_contour(
    aes(z = fit, color = count, linetype = count),
    binwidth = 0.1
  ) +
  scale_color_manual(values = crc_fc)
```

6.4 Further Reading

Palmer (1983) is possibly one of the first examinations of the balls/strikes count effect on the outcome of plate appearances: it is based on data from World Series games from 1974 to 1977 and features a table resembling Figures 6.1

and 6.2. Walsh (2008) calculates the run value of a ball and of a strike at every count and uses the results for ranking baseball's best fastballs, sliders, curveballs, and change-ups. Walsh (2010) shows how umpires are (perhaps unconsciously) affected by the balls/strikes count when judging pitches. In particular, he presents a scatterplot showing a very high correlation between the strike zone area and the count run value (see Figure 6.2). Allen (2009a), Allen (2009b), and Marchi (2010) illustrate so-called platoon splits (i.e. the different effectiveness against same-handed versus opposite-handed batters) for various pitch types.

6.5 Exercises

1. Run Value of Individual Pitches

(a) Calculate the run value of a ball and of a strike at any count. For 3-ball and 2-strike counts you need the value of a walk and a strikeout respectively (you can calculate them as done for other events in Chapter 5).

(b) Compare your values to the ones proposed by John Walsh in the article https://tht.fangraphs.com/searching-for-the-games-best-pitch/.

2. Length of Plate Appearances

(a) Calculate the length, in term of pitches, of the average plate appearance by batting position using Retrosheet data for the 2016 season.

(b) Does the eighth batter in the National League behave differently than his counterpart in the American League?

(c) Repeat the calculations in (a) and (b) for the 1991 and 2016 seasons and comment on any differences between the seasons that you find.

3. Pickoff Attempts
Identify the baserunners who, in the 2016 season, drew the highest number of pickoff attempts when standing at first base with second base unoccupied.

4. Umpire's Strike Zone

By drawing a contour plot, compare the umpire's strike zone for left-handed and right-handed batters. Use only the rows of the data frame where the pitch type is a four-seam fastball.

5. Umpire's Strike Zone, Continued

By drawing one or more contour plots, compare the umpire's strike zone by pitch type. For example, compare the 50/50 contour lines of four-seam fastballs and curveballs when a right-handed batter is at the plate.

7

Catcher Framing

7.1 Introduction

In this chapter we explore the idea of catcher framing ability using Statcast data from the 2022 season.

The story of catcher framing ability in sabermetrics is an interesting one. Historically, scouts and coaches insisted that certain catchers had the ability to "frame" pitches for umpires. The idea was that by holding the glove relatively still, you could trick the umpire into calling a pitch a strike even if it was technically outside of the strike zone (see Lindbergh (2013) for a great visual explanation). Sabermetricians were generally dubious about both the existence and the impact of this skill. Most people who had studied the impact of catcher defense concluded that it was not nearly as valuable as scouts and coaches believed.

Part of the problem was that until the mid-2000s, pitch-level data was hard to come by. With the advent of PITCHf/x, more sophisticated modeling techniques became viable on these more granular data. New studies that estimated the impact of catcher framing substantiated both the existence of a persistent ability (i.e., catchers with good framing numbers stayed good over time) and the magnitude of the effect (i.e., good framers were actually really valuable) (Turkenkopf 2008; Fast 2011; Brooks and Pavlidis 2014; Brooks, Pavilidis, and Judge 2015; Deshpande and Wyner 2017; Judge 2018).

These new findings led to changes in the baseball industry—defensive-minded catchers like José Molina starting getting multi-year contracts that were not justified by their batting skill. Minor league instruction placed greater emphasis on improving framing skills. Of course, as soon as MLB decides to let robots call balls and strikes, then this catcher framing ability will evaporate instantly.

This issue is a nice parable in that it illustrates how sabermetric thinking can (and does) change based on the availability of data and the sophistication of modeling techniques, as well as how the game on the field can change due to sabermetric insights.

DOI: 10.1201/9781032668239-7

7.2 Acquiring Pitch-Level Data

We wish to collect Statcast data for only the "taken" pitches where there is
either a ball or a called strike. We illustrate this process using the following
code that is not evaluated here. We start by reading in the Statcast data for
the 2022 season. See Section 12.2 for an explanation of how to acquire a year's
worth of Statcast data. Using the `mutate()` and `case_match()` functions, we
define a variable `Outcome` that recodes the `description` variable into three
categories—"ball", "swinging_strike" and "called_strike". We also define a `Home`
variable which indicates if the home team is batting and a `Count` variable
which gives the balls and strikes count.

```
sc2022 <- here::here("data_large/statcast_rds/statcast_2022.rds") |>
  read_rds()
sc2022 <- sc2022 |>
  mutate(
    Outcome = case_match(
      description,
      c("ball", "blocked_ball", "pitchout",
        "hit_by_pitch") ~ "ball",
      c("swinging_strike", "swinging_strike_blocked",
        "foul", "foul_bunt", "foul_tip",
        "hit_into_play", "missed_bunt" ) ~ "swing",
      "called_strike" ~ "called_strike"),
    Home = ifelse(inning_topbot == "Bot", 1, 0),
    Count = paste(balls, strikes, sep = "-")
  )
```

Using the `filter()` function, the `taken` data frame consists of those pitches
where there was not a batter swing, so only balls and called strikes are included.
We use the `select()` function to select the variables of interest in this dataset
and the `write_rds()` function stores the `taken` data frame in a compressed
format in the file `sc_taken_2022.rds`.

```
taken <- sc2022 |>
  filter(Outcome != "swing")
taken_select <- select(
  taken, pitch_type, release_speed,
  description, stand, p_throws, Outcome,
  plate_x, plate_z, fielder_2_1,
  pitcher, batter, Count, Home, zone
)
write_rds(
  taken_select,
```

```
    here::here("data/sc_taken_2022.rds"),
    compress = "xz"
)
```

Once this data is stored, we can read this data into R by use of the `read_rds()` function. We focus on using a random sample of 50,000 rows of this dataset extracted using the `sample_n()` function.

```
sc_taken <- read_rds(here::here("data/sc_taken_2022.rds"))
set.seed(12345)
taken <- sample_n(sc_taken, 50000)
```

7.3   Where Is the Strike Zone?

In order to understand the impact of catcher framing, we need a way to characterize the probability that any given pitch is called a strike. In the Statcast data, each pitch has an `Outcome` variable, which is `called_strike` for a called strike and `ball` for a ball. We plot these outcomes in Figure 7.1.

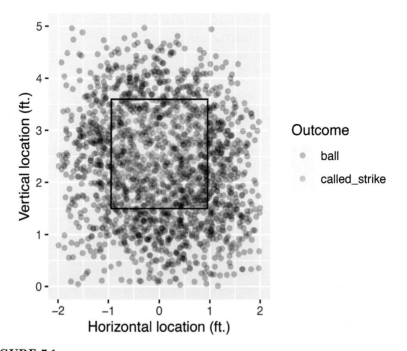

FIGURE 7.1
Scatterplot of locations of balls and called strikes for 2000 pitches, 2022 season.

Note that pitches thrown in the strike zone tend to be called a strike. Note also that many pitches are called strikes even though they are technically outside of the strike zone.

```
plate_width <- 17 + 2 * (9/pi)
k_zone_plot <- ggplot(
  NULL, aes(x = plate_x, y = plate_z)
) +
  geom_rect(
    xmin = -(plate_width/2)/12,
    xmax = (plate_width/2)/12,
    ymin = 1.5,
    ymax = 3.6, color = crcblue, alpha = 0
  ) +
  coord_equal() +
  scale_x_continuous(
    "Horizontal location (ft.)",
    limits = c(-2, 2)
  ) +
  scale_y_continuous(
    "Vertical location (ft.)",
    limits = c(0, 5)
  )
```

How do we know where the strike zone is? By the rulebook, only a part of the ball need pass over home plate in order for the pitch to be called a strike. Home plate is 17 inches wide, and the ball is 9 inches in circumference, so the outside edges of the strike zone from our point-of-view are about $\pm$ 0.947 feet. The top and bottom of the strike vary by batter, but are of comparatively less interest here. The object k_zone_plot is a blank **ggplot2** object on which we plot a random sample of 2000 rows of the Statcast data from in Figure 7.1.

```
k_zone_plot %+%
  sample_n(taken, size = 2000) +
  aes(color = Outcome) +
  geom_point(alpha = 0.2) +
  scale_color_manual(values = crc_fc)
```

Another way to think about the strike zone is in terms of zones that are pre-defined by Statcast. The strike zone itself is divided into a 3 × 3 grid, with four additional regions defined outside of the strike zone. We first compute the observed probability of a called strike in each one of those zones, as well as its boundaries. We use the quantile() function to mitigate the influence of outliers.

```
zones <- taken |>
  group_by(zone) |>
  summarize(
    N = n(),
    right_edge = min(1.5, max(plate_x)),
    left_edge = max(-1.5, min(plate_x)),
    top_edge = min(5, quantile(plate_z, 0.95, na.rm = TRUE)),
    bottom_edge = max(0, quantile(plate_z, 0.05, na.rm = TRUE)),
    strike_pct = sum(Outcome == "called_strike") / n(),
    plate_x = mean(plate_x),
    plate_z = mean(plate_z)
  )
```

In Figure 7.2, we plot each zone, along with the probability that a pitch taken in that zone will be called a strike. Note that these pre-defined zones are exclusive of those pitches "on the black".

```
library(ggrepel)
k_zone_plot %+% zones +
  geom_rect(
    aes(
      xmax = right_edge, xmin = left_edge,
      ymax = top_edge, ymin = bottom_edge,
      fill = strike_pct, alpha = strike_pct
    ),
    color = "lightgray"
  ) +
  geom_text_repel(
    size = 3,
    aes(
      label = round(strike_pct, 2),
      color = strike_pct < 0.5
    )
  ) +
  scale_fill_gradient(low = "gray70", high = crcblue) +
  scale_color_manual(values = crc_fc) +
  guides(color = FALSE, alpha = FALSE)
```

7.4 Modeling Called Strike Percentage

The zone-based strike probabilities in Figure 7.2 are limited by their discrete nature. What we really want is a model that will give us the estimated strike probability for any pitch based on its horizontal and vertical location. To

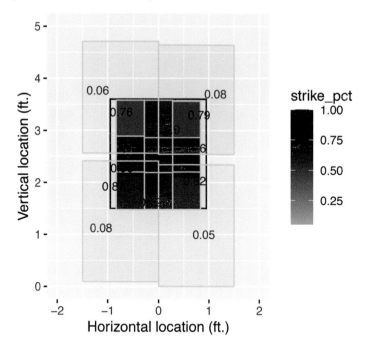

FIGURE 7.2
Strike probability for pitches taken in pre-defined areas of the strike zone.

this end, we fit a *generalized additive model*. This model will fit a smooth surface over the entire area, while including only the two explanatory variables for location. The s() function from the **mgcv** package indicates over which variables the smoothing is to occur (plate_x and plate_z). We set the family argument to binomial, to ensure that an appropriate link function (in this case, the logistic function) is used to model our binary response variable, which is defined by the Boolean expression Outcome == "called_strike".

```
library(mgcv)
strike_mod <- gam(
  Outcome == "called_strike" ~ s(plate_x, plate_z),
  family = binomial,
  data = taken
)
```

7.4.1 Visualizing the estimates

An easy way to visualize the estimates produced by our model is to plot the fitted values. Here we use the augment() function from the **broom** package to compute these fitted values and add them to our data frame. The type.predict

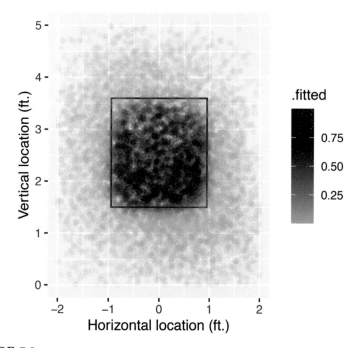

FIGURE 7.3
Estimated strike probability for taken pitches using a generalized additive model.

argument tells R to compute the estimates on the probability scale (i.e., of the response variable).

```
library(broom)
hats <- strike_mod |>
  augment(type.predict = "response")
```

Next, we can simply update our `k_zone_plot` object with this new data frame, add some points (`geom_point()`), and map the color aesthetic to the fitted values we just computed (`.fitted`). Figure 7.3 reveals that on these data, the GAM effectively mapped the pattern of balls and strikes.

```
k_zone_plot %+% sample_n(hats, 10000) +
  geom_point(aes(color = .fitted), alpha = 0.1) +
  scale_color_gradient(low = "gray70", high = crcblue)
```

7.4.2   Visualizing the estimated surface

Of course the GAM that we built is a continuous surface. One of the benefits of fitting such a model in the first place is that it allows us to estimate the probability of a called strike for *any* pitch whose location coordinates we know—not just the ones present in our training data set.

We can visualize our model as a surface by plotting the estimated probability across a fine grid of horizontal and vertical coordinate pairs. The **modelr** package has several functions, including `data_grid()` and `seq_range()` that help us create a grid of values relevant for our data.

```
library(modelr)
grid <- taken |>
  data_grid(
    plate_x = seq_range(plate_x, n = 100),
    plate_z = seq_range(plate_z, n = 100)
  )
```

Next, use the `augment()` function just as before, except this time, we specify the `newdata` argument to be the data frame of grid points that we just created. This results in a 10000 row data frame that contains the estimated called strike probability for each coordinate pair.

```
grid_hats <- strike_mod |>
  augment(type.predict = "response", newdata = grid)
```

Once again, we update our `k_zone_plot` with these new data. The `geom_tile()` function in Figure 7.4 offers a nice alternative to `geom_contour()`.

```
tile_plot <- k_zone_plot %+% grid_hats +
  geom_tile(aes(fill = .fitted), alpha = 0.7) +
  scale_fill_gradient(low = "gray92", high = crcblue)
tile_plot
```

7.4.3 Controlling for handedness

Contrary to what the rulebook states, it stands to reason that the effective strike zone may depend on with which hand the pitcher throws, and on which side of the plate the batter stands.

The resulting data frame has variables for `p_throws` and `stand` in addition to the location data encoded in `plate_x` and `plate_z`. We can now fit another GAM across these four variables. Note that the binary variables `p_throws` and `stand` are not smoothed, and are thus outside of the `s()` function in the model specification formula.

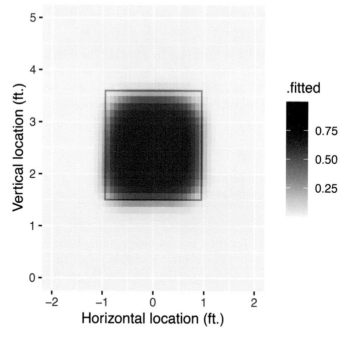

FIGURE 7.4

Estimated strike probability over a grid for taken pitches using a generalized additive model.

```
hand_mod <- gam(
  Outcome == "called_strike" ~
    p_throws + stand + s(plate_x, plate_z),
  family = binomial,
  data = taken
)
```

We must now recompute our grid of values such that they include the two additional binary variables.

```
hand_grid <- taken |>
  data_grid(
    plate_x = seq_range(plate_x, n = 100),
    plate_z = seq_range(plate_z, n = 100),
    p_throws,
    stand
  )
hand_grid_hats <- hand_mod |>
  augment(type.predict = "response", newdata = hand_grid)
```

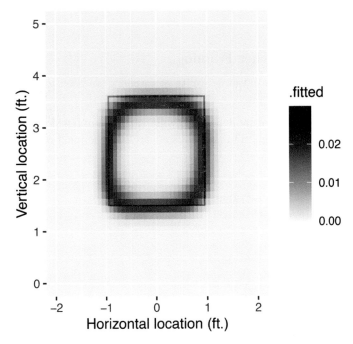

FIGURE 7.5
Standard deviation of estimated called strike probability across all four pitcher-batter handedness combinations.

The following code will produce a faceted plot across the four combinations of batter and pitcher handedness. However, as it is difficult to perceive marked difference across these four facets, we omit the plot here.

```
tile_plot %+% hand_grid_hats +
  facet_grid(p_throws ~ stand)
```

Instead, we plot the standard deviation across the four handedness combinations in Figure 7.5. In the heart of the strike zone, we see no differences due to handedness. However, the standard deviation of called strike probability is as large as 2 percentage points in some area around the perimeter of the strike zone.

```
diffs <- hand_grid_hats |>
  group_by(plate_x, plate_z) |>
  summarize(
    N = n(),
    .fitted = sd(.fitted),
    .groups = "drop"
  )
tile_plot %+% diffs
```

7.5   Modeling Catcher Framing

In order to estimate the framing ability of catchers, we need to know who the catcher is during every pitch.

To prepare these data for modeling, we will evaluate our GAM for called strike probability on each pitch. This helps us control for the location of each pitch.

```
taken <- taken |>
  filter(
    is.na(plate_x) == FALSE,
    is.na(plate_z) == FALSE
  ) |>
  mutate(
    strike_prob = predict(
      strike_mod,
      type = "response"
    )
  )
```

Next we follow Brooks, Pavilidis, and Judge (2015) in fitting a generalized linear mixed model. The response variable is whether the pitch was called a strike or a ball. Let p_j denote the probability that the jth called pitch is a strike. Our first mixed model writes the logit of the probability of a strike p_j as the sum

$$\log \frac{p_j}{1 - p_j} = \beta_0 + \beta_1 \cdot strike\_prob_j + \alpha_{c(j)}.$$

In this model, `strike_prob_j` is the "fixed effect" for the estimated called strike probability of the jth pitch based on its location computed from the previous model. So we are essentially controlling for the pitch location in this model. In addition, $\alpha_{c(j)}$ represents the effect due to the catcher $c(j)$. We assume that the individual catchers have "random" parameters, called $\alpha_1, \ldots, \alpha_C$, with mean 0 and standard deviation of s_c.

This model can be fit using the `glmer()` function in the **lme4** package. The code indicates the response variable is `Outcome == "called_strike"`, `strike_prob` is the fixed effect and `fielder_2_1` (the catcher id) represents the random effect.

```
library(lme4)
mod_a <- glmer(
  Outcome == "called_strike" ~
    strike_prob + (1|fielder_2_1),
  data = taken,
  family = binomial
)
```

We recover information about the fixed effects using the `fixed.effects()` function.

```
fixed.effects(mod_a)
```

```
(Intercept) strike_prob
      -4.00         7.67
```

Certainly different catchers will have different impacts on the probability of a called strike. The variability in these impacts is measured by the standard deviation of these random catcher effects s_c that we display by the `VarCorr()` function.

```
VarCorr(mod_a)
```

```
 Groups       Name        Std.Dev.
 fielder_2_1 (Intercept)  0.218
```

This model also provides estimates of the catcher random effects α_k that one extracts with the `ranef()` function. We put the estimates together with the catcher ids in the data frame `c_effects`.

```
c_effects <- mod_a |>
  ranef() |>
  as_tibble() |>
  transmute(
    id = as.numeric(levels(grp)),
    effect = condval
  )
```

The names of the catchers are missing, but we use the `chadwick_player_lu()` function from the **baseballr** package to construct a table for these ids and names.

```
master_id <- baseballr::chadwick_player_lu() |>
  mutate(
    mlb_name = paste(name_first, name_last),
    mlb_id = key_mlbam
  ) |>
  select(mlb_id, mlb_name) |>
  filter(!is.na(mlb_id))
```

We merge the name information with the data frame `c_effects` and display the names of the catchers with the largest and smallest random effect estimates below.

```
c_effects <- c_effects |>
  left_join(
    select(master_id, mlb_id, mlb_name),
    join_by(id == mlb_id)
  ) |>
  arrange(desc(effect))

c_effects |> slice_head(n = 6)
```

```
# A tibble: 6 x 3
      id effect mlb_name
   <dbl>  <dbl> <chr>
1 664848  0.358 Donny Sands
2 669004  0.294 MJ Melendez
3 642020  0.287 Chuckie Robinson
4 672832  0.275 Israel Pineda
5 571912  0.260 Luke Maile
6 575929  0.243 Willson Contreras

  c_effects |> slice_tail(n = 6)
```

```
# A tibble: 6 x 3
      id effect mlb_name
   <dbl>  <dbl> <chr>
1 664731 -0.293 P. J. Higgins
2 455139 -0.304 Robinson Chirinos
3 661388 -0.336 William Contreras
4 608360 -0.357 Chris Okey
5 435559 -0.357 Kurt Suzuki
6 595956 -0.390 Cam Gallagher
```

From this output, we see that Donny Sands was most effective in getting a called strike and Cam Gallagher was least effective.

One criticism of this first model is that no allowances were made for the pitcher or batter, and it is believed that both people have an impact on the probability of a called strike. We can extend the above model to include random effects for both the pitcher and the batter. We write this model as

$$\log \frac{p_j}{1 - p_j} = \beta_0 + \beta_1 strike\_prob_j + \alpha_{c(j)} + \gamma_{p(j)} + \delta_{b(j)}.$$

Here the individual pitchers are assigned parameters $\gamma_1, \ldots, \gamma_P$ that are assumed to be random from a distribution with standard deviation s_p. In addition, the individual batters are assigned parameters $\delta_1, \ldots, \delta_B$ that come from a distribution with standard deviation s_b.

This larger model is fit with a second application of the `glmer()` function, adding `batter` and `pitcher` as inputs in the regression expression.

```
mod_b <- glmer(
  Outcome == "called_strike" ~ strike_prob +
    (1|fielder_2_1) +
    (1|batter) + (1|pitcher),
  data = taken,
  family = binomial
)
```

Using the `VarCorr()` function, we display estimates of the three standard deviations s_c, s_p, and s_b. Note that the value of s_c is slightly different than it was in the previous model.

```
VarCorr(mod_b)
```

Groups	Name	Std.Dev.
pitcher	(Intercept)	0.267
batter	(Intercept)	0.251
fielder_2_1	(Intercept)	0.209

This table is helpful in identifying the components that contribute most to the total variability in called strikes. The largest standard deviations are $s_p = 0.267$ and $s_b = 0.251$ which indicate that called strikes are most influenced by the identities of the pitcher and batter, followed by the identity the catcher.

As before, we extract the catcher effect estimates by the `ranef()` function, create a data frame of ids, names, and estimates for all catchers, and then display the best and worst catchers with respect to framing. These lists are not similar to the lists prepared with the simpler random effects model, suggesting these catchers worked with different pitchers and batters who impacted the called strikes.

```
c_effects <- mod_b |>
  ranef() |>
  as_tibble() |>
  filter(grpvar == "fielder_2_1") |>
  transmute(
    id = as.numeric(as.character(grp)),
    effect = condval
  )
c_effects <- c_effects |>
  left_join(
    select(master_id, mlb_id, mlb_name),
    join_by(id == mlb_id)
```

```
  ) |>
  arrange(desc(effect))

 c_effects |> slice_head(n = 6)
```

```
# A tibble: 6 x 3
      id effect mlb_name
   <dbl>  <dbl> <chr>
1 624431  0.313 Jose Trevino
2 669221  0.277 Sean Murphy
3 425877  0.263 Yadier Molina
4 664874  0.253 Seby Zavala
5 543309  0.229 Kyle Higashioka
6 608700  0.221 Kevin Plawecki
```

```
 c_effects |> slice_tail(n = 6)
```

```
# A tibble: 6 x 3
      id effect mlb_name
   <dbl>  <dbl> <chr>
1 596117 -0.277 Garrett Stubbs
2 435559 -0.281 Kurt Suzuki
3 521692 -0.291 Salvador Perez
4 553869 -0.327 Elias Díaz
5 455139 -0.336 Robinson Chirinos
6 669004 -0.347 MJ Melendez
```

This is clearly not a thorough analysis since we only used a small dataset and did not include other effects such as umpires that could impact the called strike probability. But these mixed models with inclusion of fixed and random effects are very useful for obtaining estimates of player abilities making adjustments for other relevant inputs.

7.6 Further Reading

The first study of catcher framing using PITCHf/x was Turkenkopf (2008). See Fast (2011) for a follow-up piece. Lindbergh (2013) provides a highly-readable lay overview of the evolution of thinking on catcher framing. More sophisticated models for catcher framing include Brooks and Pavlidis (2014), Brooks, Pavilidis, and Judge (2015), Judge (2018), Deshpande and Wyner (2017).

7.7 Exercises

1. Strike Probabilities on a Grid

a. Divide the zone region into bins by use of the following code.

```
seq_x <- seq(-1.4, 1.4, by = 0.4)
seq_z <- seq(1.1, 3.9, by = 0.4)
taken <- taken |>
  mutate(
    plate_x = cut(plate_x, seq_x),
    plate_z = cut(plate_z, seq_z)
  )
```

b. By use of the `group_by()` and `summarize()` functions, find the number of strikes and balls among called pitches in each bin.
c. Find the percentage of strikes in each bin. Comment on any interesting patterns in these strike percentages across bins.

2. Strike Probability Batter Effects

In the first exercise, the strike probability percentages were found for different zones. By tabulating the balls and strikes across bins and for the variable `stand`, explore how the strike probabilities vary by the side of the batter.

3. Strike Probability Pitcher Effects

In the first exercise, the strike probability percentages were found for different zones. By tabulating the balls and strikes across bins and for the variable `p_throws`, explore how the strike probabilities vary by the throwing arm of the pitcher.

4. Count Effects

One way to explore the effect of the count on a strike probability is to fit the logistic model using the `glm()` function:

```
fit <- glm(
  Outcome == "called_strike" ~ Count,
  data = taken, family = binomial
)
```

In this expression, `Count` is a new variable derived from the `balls` and `strikes` variables in the `taken` data frame. From the output of this fit, interpret how the strike probability depends on the count.

5. Home/Away Effects

One way to explore the effect of home field on a strike probability is to fit the logistic model using the `glm()` function:

```
fit <- glm(
  Outcome == "called_strike" ~ Home,
  data = taken, family = binomial
)
```

In this expression, `Home` is a new variable that is equal to one if the batter is from the home team, and equal to zero otherwise. From the output of this fit, interpret how the strike varies among home and away batters.

8

Career Trajectories

8.1 Introduction

The R system is well-suited for fitting statistical models to data. One popular topic in sabermetrics is the rise and fall of a player's season batting, fielding, or pitching statistics from his MLB debut to retirement. Generally, it is believed that most players peak in their late 20s, although some players tend to peak at later ages. A simple way of modeling a player's trajectory is by means of a quadratic or parabolic curve. Using the `lm()` (linear model) function in R, it is straightforward to fit this model using the player's age and his OPS statistics.

We begin in Section 8.2 by considering a famous career trajectory. Mickey Mantle made an immediate impact on the New York Yankees at age 19 and quickly matured into one of the best hitters in baseball. But injuries took a toll on Mantle's performance and his hitting declined until his retirement at age 36. We use Mantle to introduce the quadratic model. Using this model, one can define his peak age, maximum performance, and the rate of improvement and decline in performance.

To compare career performances of similar players, it is helpful to contrast their trajectories, and Section 8.3 illustrates the computation of many fitted trajectories. Using Bill James' notion of similarity scores, we write a function that finds players who are most similar to a given hitter. Then we graphically compare the OPS trajectories of these similar players; by viewing these graphs we gain a general understanding of the possible trajectory shapes.

A general problem focuses on a player's peak age. In Section 8.4, we look at the fitted trajectories of all hitters with at least 2000 career at-bats. The pattern of peak ages across eras and as a function of the number of career at-bats is explored. Also, since it is common to compare players who play the same position, in Section 8.5 we focus on the period 1985–1995 and contrast the peak ages for players who play different fielding positions.

DOI: 10.1201/9781032668239-8

8.2   Mickey Mantle's Batting Trajectory

To start looking at career trajectories, we consider batting data from the great slugger Mickey Mantle. To obtain his season-by-season hitting statistics, we load the **Lahman** package, which includes the data frames `People` and `Batting`. In addition, we load the **tidyverse** package.

```
library(tidyverse)
library(Lahman)
```

We first extract Mantle's `playerID` from the `People` data frame. Using the `filter()` function, we find the line in the `People` data file where `nameFirst` equals "Mickey" and `nameLast` equals "Mantle". His player id is stored in the vector `mantle_id`.

```
mantle_id <- People |>
  filter(nameFirst == "Mickey", nameLast == "Mantle") |>
  pull(playerID)
```

One small complication is that certain statistics such as `SF` and `HBP` were not recorded for older seasons and are currently coded as `NA`. A convenient way of recoding these missing values to 0 is by the `replace_na()` function from the **tidyr** package.

```
batting <- Batting |>
  replace_na(list(SF = 0, HBP = 0))
```

To compute Mantle's age for each season, we need to know his birth year, which is available in the `People` data frame. Major League Baseball defines a player's age as his age on June 30 of that particular season.

We obtain Mantle's batting statistics by means of the user-defined function `get_stats()`. The input is the player id and the output is a data frame containing the player's hitting statistics. This function computes the player's age (variable `Age`) for all seasons, and also computes the player's slugging percentage (`SLG`), on-base percentage (`OBP`), and OPS for all seasons.

```
get_stats <- function(player_id) {
  batting |>
    filter(playerID == player_id) |>
    inner_join(People, by = "playerID") |>
    mutate(
```

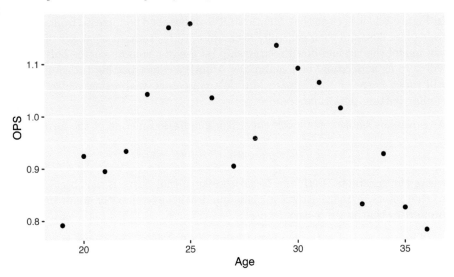

FIGURE 8.1
Scatterplot of OPS against age for Mickey Mantle.

```
    birthyear = if_else(
      birthMonth >= 7, birthYear + 1, birthYear
    ),
    Age = yearID - birthyear,
    SLG = (H - X2B - X3B - HR + 2 * X2B + 3 * X3B + 4 * HR) / AB,
    OBP = (H + BB + HBP) / (AB + BB + HBP + SF),
    OPS = SLG + OBP
  ) |>
    select(Age, SLG, OBP, OPS)
}
```

After reading the function **get_stats()** into R, we obtain Mantle's statistics
by applying this function with input **mantle_id**. The resulting data frame of
hitting statistics is stored in **Mantle**.

```
Mantle <- get_stats(mantle_id)
```

A good measure of batting performance is OPS, the sum of a player's slugging
percentage and his on-base percentage. How do Mantle's OPS season values
vary as a function of his age? To address this question, we use **ggplot2** to
construct a scatterplot of OPS against age (see Figure 8.1).

```
ggplot(Mantle, aes(Age, OPS)) + geom_point()
```

In Figure 8.1, it is clear that Mantle's OPS values tend to increase from age 19 to his late 20s, and then generally decrease until his retirement at age 36. One can model this up-and-down relationship by use of a smooth curve. This curve will help us understand and summarize Mantle's career batting trajectory and will make it easier to compare Mantle's trajectory with other players with similar batting performances.

A convenient choice of smooth curve is a quadratic function of the form

$$A + B(Age - 30) + C(Age - 30)^2,$$

where the constants A, B, and C are chosen so that curve is the "best" match to the points in the scatterplot. This quadratic curve has the following nice properties that make it easy to use.

1. The constant A is the predicted value of OPS when the player is 30 years old.

2. The function reaches its largest value at

$$PEAK\_AGE = 30 - \frac{B}{2C}.$$

   This is the age where the player is estimated to have his peak batting performance during his career.

3. The maximum value of the curve is

$$MAX = A - \frac{B^2}{4C}.$$

   This is the estimated largest OPS of the player over his career.

4. The coefficient C, typically a negative value, tells us about the degree of curvature in the quadratic function. If a player has a "large" value of C, this indicates that he more rapidly reaches his peak level and more rapidly decreases in ability until retirement. One simple interpretation is that C represents the change in OPS from his peak age to one year later.

We write a new function `fit_model()` to fit this quadratic curve to a player's batting data. The input to this function is a data frame `d` containing the player's batting statistics including the variables `Age` and `OPS`. The function `lm()` is used to fit the quadratic curve. The formula

$$OPS \sim I(Age - 30) + I((Age - 30)^2)$$

indicates that `OPS` is the response and (`Age` - 30) and (`Age` - 30)^2 are the predictors. The estimated coefficients A, B, and C are stored in the vector `b` using the `coef()` function. The peak age and maximum value are stored in the variables `Age_max` and `Max`.

```
fit_model <- function(d) {
  fit <- lm(OPS ~ I(Age - 30) + I((Age - 30)^2), data = d)
  b <- coef(fit)
  Age_max <- 30 - b[2] / b[3] / 2
  Max <- b[1] - b[2] ^ 2 / b[3] / 4
  list(fit = fit, Age_max = Age_max, Max = Max)
}
```

We then apply the function `fit_model()` to Mantle's data frame—the output of this function F2 includes the object that stores all of the calculations of the quadratic fit. In addition, this function outputs the peak age and maximum value displayed in the following code.

```
F2 <- fit_model(Mantle)
F2 |>
  pluck("fit") |>
  coef()
```

```
(Intercept)    I(Age - 30) I((Age - 30)^2)
   1.04313       -0.02288        -0.00387
```

```
c(F2$Age_max, F2$Max)
```

```
I(Age - 30) (Intercept)
      27.04        1.08
```

The best fitting curve is given by

$$1.04313 - 0.02288(Age - 30) - 0.00387(Age - 30)^2.$$

Using this model, Mantle peaked at age 27 and his maximum OPS for the curve is estimated to be 1.08. The estimated value of the curvature parameter is -0.00387; thus Mantle's decrease in OPS between his peak age and one year older is 0.00387.

We place this best quadratic curve on the scatterplot. The `geom_smooth()` function is used to estimate Mantle's OPS from the curve for the sequence of age values and overlay these values as a line on the current plot. Applications of `geom_vline()` and `geom_hline()` show the locations of the peak age and the maximum, respectively, and the `annotate()` function is used to label these values. The resulting graph is displayed in Figure 8.2.

```
ggplot(Mantle, aes(Age, OPS)) + geom_point() +
  geom_smooth(
    method = "lm", se = FALSE, linewidth = 1.5,
    formula = y ~ poly(x, 2, raw = TRUE)
```

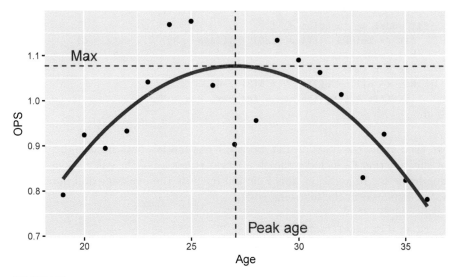

FIGURE 8.2
Career trajectory of OPS measure for Mickey Mantle with the peak age and maximum OPS identified from the quadratic smoothing curve.

```
) +
geom_vline(
  xintercept = F2$Age_max,
  linetype = "dashed", color = "red"
) +
geom_hline(
  yintercept = F2$Max,
  linetype = "dashed", color = "red"
) +
annotate(
  geom = "text", x = c(29, 20), y = c(0.72, 1.1),
  label = c("Peak age", "Max"), size = 5,
  color = "red"
)
```

Although the focus was on the best fitting quadratic curve, more details about the fitting procedure are stored in the output of `lm()` and in the variable F2. We display part of the output by finding the `summary()` of the fit. Here, we illustrate the use of the `pluck()` function to retrieve items from a list.

```
F2 |> pluck("fit") |> summary()
```

```
Call:
```

```
lm(formula = OPS ~ I(Age - 30) + I((Age - 30)^2), data = d)

Residuals:
    Min      1Q  Median      3Q     Max
-0.1728 -0.0401  0.0220  0.0451  0.1282

Coefficients:
                 Estimate Std. Error t value Pr(>|t|)
(Intercept)      1.043134   0.027901   37.39  3.2e-16 ***
I(Age - 30)     -0.022883   0.005638   -4.06    1e-03 **
I((Age - 30)^2) -0.003869   0.000828   -4.67    3e-04 ***
---
Signif. codes:  0 '***' 0.001 '**' 0.01 '*' 0.05 '.' 0.1 ' ' 1

Residual standard error: 0.0842 on 15 degrees of freedom
Multiple R-squared:  0.602, Adjusted R-squared:  0.549
F-statistic: 11.3 on 2 and 15 DF,  p-value: 0.001
```

The value of R^2 is 0.602; this means that approximately 60% of the variability in Mantle's OPS values can be explained by the quadratic curve. The residual standard error is equal to 0.084. Approximately 2/3 of the vertical deviations (the residuals) from the curve fall between plus and minus one residual standard error. In this case, the interpretation is that approximately 2/3 of the residuals fall between –0.084 and 0.084.

8.3 Comparing Trajectories

8.3.1 Some preliminary work

When we think about hitting trajectories of players, one relevant variable seems to be a player's fielding position. Hitting expectations of a catcher—an important defensive position—are different from the hitting expectations of a first baseman. To compare trajectories of players with the same position, fielding position should be recorded in our database. Recall that the data frame batting has already been created. The fielding data is stored in the Fielding data frame from the **Lahman** package.

Many players in the history of baseball have had short careers and in our study of trajectories, it seems reasonable to limit our analysis to players who have had a minimum number of at-bats. We consider only players with 2000 at-bats; this will remove hitting data of pitchers and other players with short careers. To take this subset of the batting data frame, we use the group_by() and summarize() functions in the **dplyr** package to compute the career at-bats for all players; the new variable is called `AB_career`. Using the `inner_join()` function, we add this new variable to the `batting` data frame. Finally, using the `filter()` function, we create a new data frame `batting_2000` consisting only of the "minimum 2000 AB" hitters.

```
batting_2000 <- batting |>
  group_by(playerID) |>
  summarize(AB_career = sum(AB, na.rm = TRUE)) |>
  inner_join(batting, by = "playerID") |>
  filter(AB_career >= 2000)
```

To add fielding information to our data frame, we need to find the primary fielding position for a given player. We tally the number of games played at each possible position, and the data frame `Positions` returns for each player the position where he played the most games.[1]

```
Positions <- Fielding |>
  group_by(playerID, POS) |>
  summarize(Games = sum(G)) |>
  arrange(playerID, desc(Games)) |>
  filter(POS == first(POS))
```

We then combine this new fielding information with the `batting_2000` data frame using the `inner_join()` function.

```
batting_2000 <- batting_2000 |>
  inner_join(Positions, by = "playerID")
```

8.3.2   Computing career statistics

We will find groups of similar hitters on the basis of their career statistics. Toward this goal, one needs to compute the career games played, at-bats, runs, hits, etc., for each player in the `batting_2000` data frame. This is conveniently done using the `group_by()` and `summarize()` functions. In the R code, we use the `across()` function to find the sum of a collection of different batting statistics, defined in the vector `vars`, for each hitter. We create a new data frame `C_totals` with the player id variable `playerID` and new career variables.

```
my_vars <- c("G", "AB", "R", "H", "X2B", "X3B",
             "HR", "RBI", "BB", "SO", "SB")
```

---

[1]In the rare case where there are two or more positions with the most games played, the function `first()` will take the first position.

```
C_totals <- batting |>
  group_by(playerID) |>
  summarize(across(all_of(my_vars), ~sum(.x, na.rm = TRUE)))
```

In the new data frame, we compute each player's career batting average AVG
and his career slugging percentage SLG using `mutate()`.

```
C_totals <- C_totals |>
  mutate(
    AVG = H / AB,
    SLG = (H - X2B - X3B - HR + 2 * X2B + 3 * X3B + 4 * HR) / AB
  )
```

We then merge the career statistics data frame `C_totals` with the fielding
data frame `Positions`. Each fielding position has an associated value, and the
`case_when()` function is used to define a value `Value_POS` for each position
POS. These values were introduced by Bill James in James (1994) and displayed
in Baseball-Reference's Similarity Scores page.

```
C_totals <- C_totals |>
  inner_join(Positions, by = "playerID") |>
  mutate(
    Value_POS = case_when(
      POS == "C"  ~ 240,
      POS == "SS" ~ 168,
      POS == "2B" ~ 132,
      POS == "3B" ~ 84,
      POS == "OF" ~ 48,
      POS == "1B" ~ 12,
      TRUE ~ 0
    )
  )
```

8.3.3 Computing similarity scores

Bill James introduced the concept of similarity scores to facilitate the comparison of players on the basis of career statistics. To compare two hitters,
one starts at 1000 points and subtracts points based on the differences in
different statistical categories. One point is subtracted for each of the following
differences: (1) 20 games played, (2) 75 at-bats, (3) 10 runs scored, (4) 15 hits,
(5) 5 doubles, (6) 4 triples, (7) 2 home runs, (8) 10 runs batted in, (9) 25 walks,
(10) 150 strikeouts, (11) 20 stolen bases, (12) 0.001 in batting average, and
(13) 0.002 in slugging percentage. In addition, one subtracts the absolute value
of the difference between the fielding position values of the two players.

The function similar() will find the players most similar to a given player using similarity scores on career statistics and fielding position. One inputs the id for the particular player and the number of similar players to be found (including the given player). The output is a data frame of player statistics, ordered in decreasing order by similarity scores.

```
similar <- function(p, number = 10) {
  P <- C_totals |>
    filter(playerID == p)
  C_totals |>
    mutate(
      sim_score = 1000 -
        floor(abs(G - P$G) / 20) -
        floor(abs(AB - P$AB) / 75) -
        floor(abs(R - P$R) / 10) -
        floor(abs(H - P$H) / 15) -
        floor(abs(X2B - P$X2B) / 5) -
        floor(abs(X3B - P$X3B) / 4) -
        floor(abs(HR - P$HR) / 2) -
        floor(abs(RBI - P$RBI) / 10) -
        floor(abs(BB - P$BB) / 25) -
        floor(abs(SO - P$SO) / 150) -
        floor(abs(SB - P$SB) / 20) -
        floor(abs(AVG - P$AVG) / 0.001) -
        floor(abs(SLG - P$SLG) / 0.002) -
        abs(Value_POS - P$Value_POS)
    ) |>
    arrange(desc(sim_score)) |>
    slice_head(n = number)
}
```

To illustrate the use of this function, suppose one is interested in finding the five players who are most similar to Mickey Mantle. Recall that the player id for Mantle is stored in the vector mantle_id. We use the function similar() with inputs mantle_id and 6.

```
similar(mantle_id, 6)
```

```
# A tibble: 6 x 18
   playerID     G    AB     R     H   X2B   X3B    HR   RBI    BB
   <chr>    <int> <int> <int> <int> <int> <int> <int> <int> <int>
1 mantlmi~  2401  8102  1677  2415   344    72   536  1509  1733
2 thomafr~  2322  8199  1494  2468   495    12   521  1704  1667
3 matheed~  2391  8537  1509  2315   354    72   512  1453  1444
4 schmimi~  2404  8352  1506  2234   408    59   548  1595  1507
5 sheffga~  2576  9217  1636  2689   467    27   509  1676  1475
```

```
6 sosasa01   2354  8813  1475  2408    379     45    609   1667    929
# i 8 more variables: SO <int>, SB <int>, AVG <dbl>, SLG <dbl>,
#   POS <chr>, Games <int>, Value_POS <dbl>, sim_score <dbl>
```

From reading the player ids, we see five players who are similar in terms of career hitting statistics and position: Frank Thomas, Eddie Mathews, Mike Schmidt, Gary Sheffield, and Sammy Sosa.

8.3.4 Defining age, OBP, SLG, and OPS variables

To fit and graph hitting trajectories for a group of similar hitters, one needs to have age and OPS statistics for all seasons for each player. One complication with working with the **Lahman** `Batting` table is that separate batting lines are used for batters who played with multiple teams during a season. There is a variable `stint` that gives different values (1, 2, ...) in the case of a player with multiple teams. The following code uses the `group_by()` and `summarize()` functions to combine these multiple lines into a single row for each player in each year. It also computes the batting measures `SLG`, `OBP`, and `OPS`. (Recall we had earlier replaced any missing values for `HBP` and `SF` with zeros, so there will be no missing values in the calculation of the `OBP` and `OPS` variables.)

```
batting_2000 <- batting_2000 |>
  group_by(playerID, yearID) |>
  summarize(
    G = sum(G), AB = sum(AB), R = sum(R),
    H = sum(H), X2B = sum(X2B), X3B = sum(X3B),
    HR = sum(HR), RBI = sum(RBI), SB = sum(SB),
    CS = sum(CS), BB = sum(BB), SH = sum(SH),
    SF = sum(SF), HBP = sum(HBP),
    AB_career = first(AB_career),
    POS = first(POS)
  ) |>
  mutate(
    SLG = (H - X2B - X3B - HR + 2 * X2B + 3 * X3B + 4 * HR) / AB,
    OBP = (H + BB + HBP) / (AB + BB + HBP + SF),
    OPS = SLG + OBP
  )
```

Thus, we create a new version of the `batting_2000` data frame where the hitting statistics for a player for a season are recorded on a single line.

The next task is to obtain the ages for all players for all seasons. Recall that we used a similar technique in Section 3.7 to compute the MLB birth year for a particular player. Here, we compute the birth year for all players, and use the `inner_join()` function to merge this birth year information with the batting data. Now that we have birth years for all players, we can define the new variable `Age` as the difference between the season year and the birth year.

```
batting_2000 <- batting_2000 |>
  inner_join(People, by = "playerID") |>
  mutate(
    Birthyear = if_else(
      birthMonth >= 7, birthYear + 1, birthYear
    ),
    Age = yearID - Birthyear
  )
```

A small complication is that the birth year is not recorded for a few 19th century ballplayers, and so the age variable is missing for these variables. Accordingly, we use the `drop_na()` function to omit the age records that are missing, and the updated data frame `batting_2000` only contains players for which the `Age` variable is available.

```
batting_2000 |> drop_na(Age) -> batting_2000
```

8.3.5   Fitting and plotting trajectories

Given a group of similar players, we write a function `plot_trajectories()` to fit quadratic curves to each player and graph the trajectories in a way that facilitates comparisons. This function takes as input the first and last name of the player, the number of players to compare (including the one of interest), and the number of columns in the multipanel plot.

The function `plot_trajectories()` first uses the `People` data frame to find the player id for the player. It then uses the `similar()` function to find a vector of player ids `player_list`. The data frame `Batting_new` consists of the season batting statistics for only the players in the player list. The graphing is done by use of the **ggplot2** package. The use of `geom_smooth()` with the `formula` argument of $y \sim x + I(x^2)$ constructs trajectory curves of `Age` and `Fit` for all players. The `facet_wrap()` function with the `ncol` argument places these trajectories on separate panels where the number of columns in the multipanel display is the value specified in the argument of the function.

```
plot_trajectories <- function(player, n_similar = 5, ncol) {
  flnames <- unlist(str_split(player, " "))

  player <- People |>
    filter(nameFirst == flnames[1], nameLast == flnames[2]) |>
    select(playerID)

  player_list <- player |>
    pull(playerID) |>
```

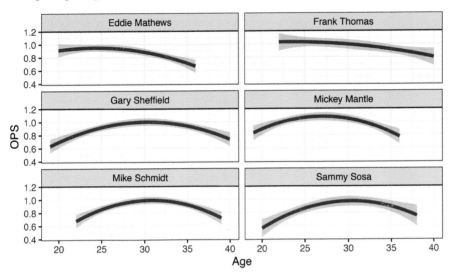

FIGURE 8.3
Estimated career trajectories for Mickey Mantle and five similar hitters.

```
        similar(n_similar) |>
        pull(playerID)

    Batting_new <- batting_2000 |>
        filter(playerID %in% player_list) |>
        mutate(Name = paste(nameFirst, nameLast))

    ggplot(Batting_new, aes(Age, OPS)) +
        geom_smooth(
            method = "lm",
            formula = y ~ x + I(x^2),
            linewidth = 1.5
        ) +
        facet_wrap(vars(Name), ncol = ncol) +
        theme_bw()
}
```

Here are several examples of the use of `plot_trajectories()`. In Figure 8.3, we compare Mickey Mantle's trajectory with those of his five most similar hitters.

```
plot_trajectories("Mickey Mantle", 6, 2)
```

We compare Derek Jeter's OPS trajectory with eight similar players in Figure 8.4. In this case note that the **ggplot2** object is saved in the variable

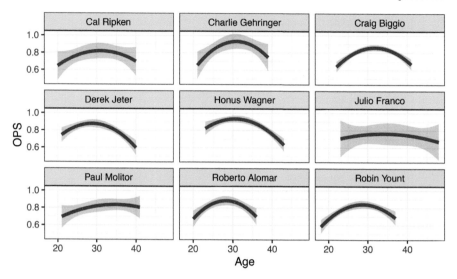

FIGURE 8.4
Estimated career trajectories for Derek Jeter and eight similar hitters.

`dj_plot`. (It will be seen shortly that one can extract the relevant data from this object.)

```
dj_plot <- plot_trajectories("Derek Jeter", 9, 3)
dj_plot
```

Looking at Figures 8.3 and 8.4, we see notable differences in these trajectories.

- There are players such as Eddie Mathews, Frank Thomas, Mickey Mantle, and Roberto Alomar who appeared to peak early in their careers.
- In contrast, players such as Mike Schmidt, Craig Biggio, and Julio Franco peaked in their 30s.
- The players also show differences in the shape of the trajectory. For example, Paul Molitor had a relatively flat trajectory while Roberto Alomar had a trajectory with high curvature.

One can summarize these trajectories by the peak age, the maximum value, and the curvature. To begin, the component `dj_plot$data` contains the batting data for Jeter and the group of similar players. We use the `group_split()` function to split the data into a list with one element for each player. Then, we use `map()` to fit a quadratic model to each player. The `tidy()` function from the **broom** package helps us recover the coefficients in a tidy fashion.

The output data frame `regressions` contains the regression estimates for each player.

```
library(broom)
data_grouped <- dj_plot$data |>
  group_by(Name)
player_names <- data_grouped |>
  group_keys() |>
  pull(Name)
regressions <- data_grouped |>
  group_split() |>
  map(~lm(OPS ~ I(Age - 30) + I((Age - 30) ^ 2), data = .)) |>
  map(tidy) |>
  set_names(player_names) |>
  bind_rows(.id = "Name")
regressions |>
  slice_head(n = 6)
```

```
# A tibble: 6 x 6
  Name              term    estimate std.error statistic  p.value
  <chr>             <chr>      <dbl>     <dbl>     <dbl>    <dbl>
1 Cal Ripken        (Inte~   0.820    0.0436      18.8   2.74e-13
2 Cal Ripken        I(Age~   0.00273  0.00479      0.570 5.76e- 1
3 Cal Ripken        I((Ag~  -0.00148  0.000887    -1.67  1.12e- 1
4 Charlie Gehringer (Inte~   0.932    0.0415      22.4   1.60e-13
5 Charlie Gehringer I(Age~   0.00507  0.00504      1.00  3.30e- 1
6 Charlie Gehringer I((Ag~  -0.00285  0.00103     -2.76  1.40e- 2
```

Next, using the `summarize()` function together with `regressions`, we find summary statistics for all players including the peak age, maximum value, and curvature. Recall this calculation is illustrated for Jeter and eight similar players.

```
S <- regressions |>
  group_by(Name) |>
  summarize(
    b1 = estimate[1],
    b2 = estimate[2],
    Curvature = estimate[3],
    Age_max = round(30 - b2 / Curvature / 2, 1),
    Max = round(b1 - b2 ^ 2 / Curvature / 4, 3)
  )
```

To help understand the differences between the nine player trajectories, we use the `ggplot()`function to construct a scatterplot of the peak ages and the curvature statistics. The `geom_label_repel()` function is used to add player labels.

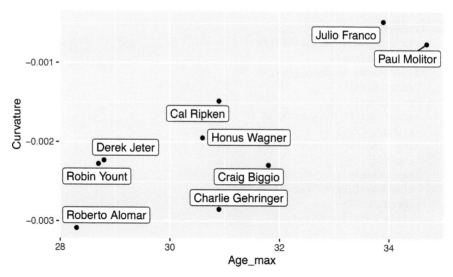

FIGURE 8.5
Estimated peak ages and curvature statistics for Derek Jeter and eight players
most similar to him.

```
library(ggrepel)
ggplot(S, aes(Age_max, Curvature, label = Name)) +
  geom_point() + geom_label_repel()
```

Figure 8.5 clearly indicates that Alomar peaked at an early age, Franco
and Molitor at a late age, and Alomar and Gehringer exhibited the greatest
curvature, indicating they rapidly declined in performance after the peak.

8.4   General Patterns of Peak Ages

8.4.1   Computing all fitted trajectories

We have explored the hitting career trajectories of groups of similar players.
How have career trajectories changed over the history of baseball? We'll focus
on a player's peak age and explore how this has changed over time. We will also
explore the relationship between peak age and the number of career at-bats.

We wish to focus on players that are no longer active, so we use the `finalgame`
variable in the `People` data frame to restrict our attention to players whose
final game was before November 1, 2021.

```
not_current_playerID <- People |>
  filter(finalGame < "2021-11-01") |>
  pull(playerID)
batting_2000 <- batting_2000 |>
  filter(playerID %in% not_current_playerID)
```

For each player, the variable `yearID` contains the seasons played. We define a
new variable `Midyear` to be the average of a player's first and last seasons. We
use the `group_by()` and `summarize()` functions to compute `Midyear` for all
players and add this new variable to the `batting_2000` data frame using the
`inner_join()` function.

```
midcareers <- batting_2000 |>
  group_by(playerID) |>
  summarize(
    Midyear = (min(yearID) + max(yearID)) / 2,
    AB_total = first(AB_career)
  )
batting_2000 <- batting_2000 |>
  inner_join(midcareers, by = "playerID")
```

Quadratic curves to all of the career trajectories are fit by another application
of the `map()` function from the **purrr** package. First, we apply `group_split()`,
where `playerID` is the grouping variable and the model fit is to be fit to each
player's data separately. The output `models` is a data frame containing the
coefficients for all players, where a row corresponds to a particular player.

```
batting_2000_grouped <- batting_2000 |>
  group_by(playerID)
ids <- batting_2000_grouped |>
  group_keys() |>
  pull(playerID)
models <- batting_2000_grouped |>
  group_split() |>
  map(~lm(OPS ~ I(Age - 30) + I((Age - 30)^2), data = .)) |>
  map(tidy) |>
  set_names(ids) |>
  bind_rows(.id = "playerID")
```

We compute the estimated peak ages for all players using the formula
$Peak\_age = 30 - B/(2C)$. We add the new variable `Peak_age` to the data
frame `beta_coefs`.

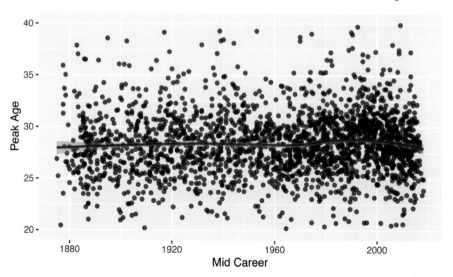

FIGURE 8.6
Scatterplot of peak age and midcareer for all players with at least 2000 career
at-bats. A smoothing curve is used to see the general pattern in the scatterplot.

```
beta_coefs <- models |>
  group_by(playerID) |>
  summarize(
    A = estimate[1],
    B = estimate[2],
    C = estimate[3]
  ) |>
  mutate(Peak_age = 30 - B / 2 / C) |>
  inner_join(midcareers, by = "playerID")
```

8.4.2   Patterns of peak age over time

To investigate how the peak age varies over the history of baseball, we construct
a scatterplot of `Peak_age` against `Midyear` by use of the `ggplot()` function.
It is difficult to see the general pattern by just looking at the scatterplot, so
we use the `geom_smooth()` function to fit a smooth curve and add it to the
plot (see Figure 8.6).

```
age_plot <- ggplot(beta_coefs, aes(Midyear, Peak_age)) +
  geom_point(alpha = 0.5) +
  geom_smooth(color = "red", method = "loess") +
  ylim(20, 40) +
```

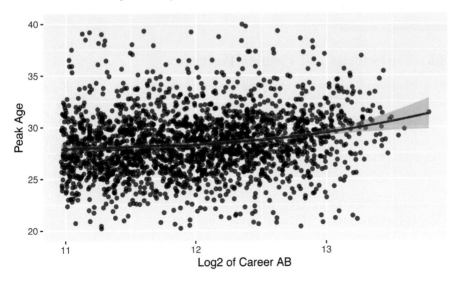

FIGURE 8.7
Scatterplot of the logarithm of the career at-bats and peak age for all players
with at least 2000 career at-bats. A smoothing curve is used to see the general
pattern in the scatterplot.

```
xlab("Mid Career") + ylab("Peak Age")
age_plot
```

In Figure 8.6, we see a gradual increase in peak age over time. The peak age for
an average player was approximately 27 in 1880 and this average has gradually
increased to 28 from 1880 to 2016.

8.4.3 Peak age and career at-bats

Is there any relationship between a player's peak age and his career at-bats?
Using **ggplot2**, we construct a graph of `Peak_age` against the logarithm (base
2) of the career at-bats variable `AB_career`. We plot the at-bats on a log scale,
so that the points are more evenly spread out over all possible values. Again
we overlay a LOESS smoothing curve to see the pattern in Figure 8.7.

```
age_plot +
  aes(x = log2(AB_total)) +
  xlab("Log2 of Career AB")
```

Here we see a clear relationship. Players with relatively short careers and
2000 career at-bats tend to peak about age 27. In contrast, players with long
careers—say 9000 or more at-bats—tend to peak at ages closer to 30.

8.5   Trajectories and Fielding Position

In comparing players, we typically want to compare players at the same fielding
position. The primary fielding position POS was already defined and we use
this variable to compare peak ages of players categorized by position.

Suppose we consider the players whose mid-career is between 1985 and 1995.
Using the filter() function, we create a new data frame Batting_2000a
consisting of only these players.

```
batting_2000a <- batting_2000 |>
  filter(Midyear >= 1985, Midyear <= 1995)
```

Another application of the map() function is used to fit quadratic curves to
the trajectory data for the players in the batting_2000a data frame and the
quadratic fits are stored in the object models. By use of the summarize()
and mutate() functions, we summarize the regression fits. The output is the
estimated coefficients A, B, C, the player's estimated peak age Peak_age and
his fielding position Position; this information is stored in the data frame
beta_estimates.

```
batting_2000a_grouped <- batting_2000a |>
  group_by(playerID)
ids <- batting_2000a_grouped |>
  group_keys() |>
  pull(playerID)
models <- batting_2000a_grouped |>
  group_split() |>
  map(~lm(OPS ~ I(Age - 30) + I((Age - 30)^2), data = .)) |>
  map(tidy) |>
  set_names(ids) |>
  bind_rows(.id = "playerID")

beta_estimates <- models |>
  group_by(playerID) |>
  summarize(
    A = estimate[1],
    B = estimate[2],
    C = estimate[3]
  ) |>
  mutate(Peak_age = 30 - B / 2 / C) |>
  inner_join(midcareers) |>
  inner_join(Positions) |>
  rename(Position = POS)
```

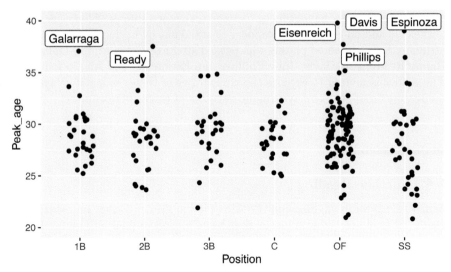

FIGURE 8.8
Scatterplot of peak ages against primary fielding position.

We focus on the primary fielding positions excluding pitcher and designated hitter. The `filter()` function removes these other positions. We combine the trajectory and fielding information with the `People` info by use of the `inner_join()` function and store the combined information in the data frame `beta_fielders`.

```
beta_fielders <- beta_estimates |>
  filter(
    Position %in% c("1B", "2B", "3B", "SS", "C", "OF")
  ) |>
  inner_join(People)
```

We use a stripchart to graph the peak ages of the players against the fielding position (see Figure 8.8). Since some of the peak age estimates are not reasonable values, the limits on the horizontal axis are set to 20 and 40.

```
ggplot(beta_fielders, aes(Position, Peak_age)) +
  geom_jitter(width = 0.2) + ylim(20, 40) +
  geom_label_repel(
    data = filter(beta_fielders, Peak_age > 37),
    aes(Position, Peak_age, label = nameLast)
  )
```

Generally, for all fielding positions, the peak ages for these 1990 players tend to fall between 27 and 32. The variability in the peak age estimates reflects the fact that hitters have different career trajectory shapes. There are three outfielders and no catchers who appear to stand out by having a high peak age estimate. Six highlighted players who peaked after age 37 are Andrés Galarraga, Randy Ready, Eric Davis, Tony Phillips, Jim Eisenreich, and Alvaro Espinoza. The reader is invited to explore the trajectories of these "unusual" players to see if they do appear to have unique patterns of career performance.

8.6 Further Reading

James (1982) wrote an essay on "Looking for the Prime". Based on a statistical study, he came to the conclusion that batters tend to peak at age 27. Berry, Reese, and Larkey (1999) give a general discussion of career trajectories of athletes from hockey, baseball, and golf. Chapter 11 of Albert and Bennett (2003) considers the career trajectories of the home run rates of nine great historical sluggers. Albert (2002) and Albert (2009) discuss general patterns of trajectories of hitters and pitchers in baseball history, and Fair (2008) performs an extensive analysis of baseball career trajectories based on quadratic models. Albert and Rizzo (2012), Chapter 7, give illustrations of regression modeling using R.

8.7 Exercises

1. Career Trajectory of Willie Mays

a. Use the `gets_stats()` function to extract the hitting data for Willie Mays for all of his seasons in his career.
b. Construct a scatterplot of Mays' OPS season values against his age.
c. Fit a quadratic function to Mays' career trajectory. Based on this model, estimate Mays' peak age and his estimated largest OPS value based on the fit.

2. Comparing Trajectories

a. Using James' similarity score measure (function `similar()`), find the five hitters with hitting statistics most similar to Willie Mays.
b. Fit quadratic functions to the (Age, OPS) data for Mays and the five similar hitters. Display the six fitted trajectories on a single panel.
c. Based on your graph, describe the differences between the six player trajectories. Which player had the smallest peak age?

3. Comparing Trajectories of the Career Hits Leaders

a. Find the batters who have had at least 3200 career hits.

b. Fit the quadratic functions to the (Age, AVG) data for this group of hitters, where AVG is the batting average. Display the fitted trajectories on a single panel.

c. On the basis of your work, which player was the most consistent hitter on average? Explain how you measured consistency on the basis of the fitted trajectory.

4. Comparing Trajectories of Home Run Hitters

a. Find the ten players in baseball history who have had the most career home runs.

b. Fit the quadratic functions to the home run rates of the ten players, where $HR_rate = HR/AB$. Display the fitted trajectories on a single panel.

c. On the basis of your work, which player had the highest estimated home run rate at his peak? Which player among the ten had the smallest peak home run rate?

d. Do any of the players have unusual career trajectory shapes? Is there any possible explanation for these unusual shapes?

5. Peak Ages in the History of Baseball

a. Find all the players who entered baseball between 1940 and 1945 with at least 2000 career at-bats.

b. Find all the players who entered baseball between 1970 and 1975 with at least 2000 career at-bats.

c. By fitting quadratic functions to the (Age, OPS) data, estimate the peak ages for all players in parts (a) and (b).

d. By comparing the peak ages of the 1940s players with the peak ages of the 1970s players, can you make any conclusions about how the peak ages have changed in this 30-year period?

9

Simulation

9.1 Introduction

A baseball season consists of a collection of games between teams, where each game consists of nine innings, and a half-inning consists of a sequence of plate appearances. Because of this clean structure, the sport can be represented by relatively simple probability models. Simulations from these models are helpful in understanding different characteristics of the game.

One attractive aspect of the R system is its ability to simulate from a wide variety of probability distributions. In this chapter, we illustrate the use of R functions to simulate a game consisting of a large number of plate appearances. Also, We use R simulate the game-to-game competition of teams during an entire season.

Section 9.2 focuses on simulating the events in a baseball half-inning using a special probability model called a *Markov chain*. The runners on base and the number of outs define a state and this probability model describes movements between states until one reaches three outs. The movement or transition probabilities are found using actual data from the 2016 season. By simulating many half-innings using this model, one gets a basic understanding of the pattern of run scoring.

Section 9.3 describes a simulation of an entire baseball season using the Bradley-Terry probability model. Teams are assigned talents from a bell-shaped (normal) distribution and a season of baseball games is played using win probabilities based on the talents. By simulating many seasons, one learns about the relationship between a team's talent and its performance in a 162-game season. We describe simulating the post-season series and assess the probability that the "best" team, that is, the team with the best ability actually wins the World Series.

DOI: 10.1201/9781032668239-9

9.2   Simulating a Half Inning

9.2.1   Markov chains

A Markov chain is a special type of probability model useful for describing movement between locations, called *states*. In the baseball context, a state is viewed as a description of the runners on base and the number of outs in an inning. Each of the three bases can be occupied by a runner or not, and so there are $2 \times 2 \times 2 = 8$ possible runner situations. Since there are three possible numbers of outs (0, 1, or 2), there are $8 \times 3 = 24$ possible runner and outs states. If we include the 3 outs state, there are a total of 25 possible states during a half-inning of baseball (see Section 5.1).

In a Markov chain, a matrix of *transition probabilities* is used to describe how one moves between the different states. For example, suppose that there are currently runners on first and second with one out. Based on the outcome of the plate appearance, the state can change. For example, the batter may hit a single; the runner on second scores and the runner on first moves to third. In this case, the new state is runners on first and third with one out. Or maybe the batter will strike out, and the new state is runners on first and second with two outs. By looking at a specific row in the transition probability matrix, one learns about the probability of moving to first and third with one out, or moving to first and second with two outs, or any other possible state.

In a Markov chain, there are two types of states: transition states and absorbing states. Once one moves into an absorbing state, one remains there and can't return to other transition states. In a half-inning of baseball, since the inning is over when there are 3 outs, this 3-outs state acts as an absorbing state.

There are some special assumptions in a Markov chain model. We assume that the probability of moving to a new state only depends on the current state. So any baseball events that happened before the current runners and outs situation are not relevant in finding the probabilities[1]. In other words, this model assumes there is not a momentum effect in batting through an inning. Also we are assuming that the probabilities of these movements are the same for all teams, against all pitchers, and for all innings during a game. Clearly, this assumption that all teams are average is not realistic, but we will address this issue in one of the other sections of this chapter.

There are several attractive aspects of using a Markov chain to model a half-inning of baseball. First, the construction of the transition probability matrix is easily done with 2016 season data using computations from Chapter 5. One can use the model to play many half-innings of baseball, and the run scoring patterns that are found resemble the actual run scoring of actual MLB

---

[1]This is often known as the *memoryless property*.

baseball. Last, there are special properties of Markov chains that simplify some interesting calculations, such as the number of players who come to bat during an inning.

9.2.2 Review of work in run expectancy

To construct the transition matrix for the Markov chain, one needs to know the frequencies of transitions from the different runners/outs states to other possible runners/outs states. One can obtain these frequencies using the Retrosheet play-by-play data from a particular season. Here we review the work from Chapter 5.

We begin by reading in play-by-play data for the 2016 season, creating the data frame `retro2016`.

```
library(tidyverse)
retro2016 <- read_rds(here::here("data/retro2016.rds"))
```

First, we use the `retrosheet_add_states()` function that we wrote in Chapter 5 and stored in the **abdwr3edata** package. This function adds a number of useful new variables to `retro2016`. Recall that in particular, we now have a new variable `state` (which gives the runner locations and the number of outs at the beginning of each play), and a another new variable `new_state` (which contains the same information at the conclusion of the play).

```
library(abdwr3edata)
retro2016 <- retro2016 |>
  retrosheet_add_states()
```

Next, we create the variable `half_inning_id` as a unique identifier for each half-inning in each baseball game. The new variable `runs` gives the number of runs scored in each play. The new data frame `half_innings` contains data aggregated over each half-inning of baseball played in 2016.

```
half_innings <- retro2016 |>
  mutate(
    runs = away_score_ct + home_score_ct,
    half_inning_id = paste(game_id, inn_ct, bat_home_id)
  ) |>
  group_by(half_inning_id) |>
  summarize(
    outs_inning = sum(event_outs_ct),
    runs_inning = sum(runs_scored),
    runs_start = first(runs),
    max_runs = runs_inning + runs_start
```

```
)
```

By using the `filter()` function from the **dplyr** package, we focus on plays where there is a change in the state or in the number of runs scored. By another application of `filter()`, we restrict attention to complete innings where there are three outs and where there is a batting event; the new dataset is called `retro2016_complete`. Here non-batting plays such as steals, caught stealing, wild pitches, and passed balls are ignored. There is obviously some consequence of removing these non-batting plays from the viewpoint of run production, and this issue is discussed later in this chapter.

```
retro2016_complete <- retro2016 |>
  mutate(
    half_inning_id = paste(game_id, inn_ct, bat_home_id)
  ) |>
  inner_join(half_innings, join_by(half_inning_id)) |>
  filter(state != new_state | runs_scored > 0) |>
  filter(outs_inning == 3, bat_event_fl)
```

In our definition of the `new_state` variable, we recorded the runner locations when there were three outs. The runner locations don't matter, so we recode `new_state` to always have the value 3 when the number of outs is equal to 3. The `str_replace()` function replaces the regular expression [0-1]{3} 3— which matches any three character binary string followed by a space and a 3—with 3.

```
retro2016_complete <- retro2016_complete |>
  mutate(new_state = str_replace(new_state, "[0-1]{3} 3", "3"))
```

9.2.3 Computing the transition probabilities

Now that the `state` and `new_state` variables are defined, one can compute the frequencies of all possible transitions between states using the `table()` function. The matrix of counts is `T_matrix`. There are 24 possible values of the beginning state `state`, and 25 values of the final state `new_state` including the 3-outs state.

```
T_matrix <- retro2016_complete |>
  select(state, new_state) |>
  table()
dim(T_matrix)
```

```
[1] 24 25
```

This matrix can be converted to a probability matrix by use of the `prop.table()` function. The resulting matrix is denoted by `P_matrix`.

```
P_matrix <- prop.table(T_matrix, 1)
dim(P_matrix)
```

```
[1] 24 25
```

Finally, we add a row to this transition probability matrix corresponding to transitions from the 3-out state. When the inning reaches 3 outs, then it stays at 3 outs, so the probability of staying in this state is 1.

```
P_matrix <- P_matrix |>
  rbind("3" = c(rep(0, 24), 1))
```

The matrix `P_matrix` now has two important properties that allow it to model transitions between states in a Markov chain: 1) it is square, and; 2) the entries in each of its rows sum to 1.

```
dim(P_matrix)
```

```
[1] 25 25
```

```
P_matrix |>
  apply(MARGIN = 1, FUN = sum)
```

```
000 0 000 1 000 2 001 0 001 1 001 2 010 0 010 1 010 2 011 0
    1     1     1     1     1     1     1     1     1     1
011 1 011 2 100 0 100 1 100 2 101 0 101 1 101 2 110 0 110 1
    1     1     1     1     1     1     1     1     1     1
110 2 111 0 111 1 111 2     3
    1     1     1     1     1
```

To better understand this transition matrix, we display the transition probabilities starting at the "000 0" state, no runners and no outs below. (Only the positive probabilities are shown and the `as_tibble()` and `pivot_longer()` functions are used to display the probabilities vertically.) The most likely transitions are to the "no runners, one out" state with probability 0.676 and to the "runner on first, no outs" state with probability 0.235. The probability of moving from the "000 0" state to the "000 0" state is 0.033; in other words, the chance of a home run with no runners on with no outs is 0.033.

```
P_matrix |>
  as_tibble(rownames = "state") |>
  filter(state == "000 0") |>
```

```
    pivot_longer(
      cols = -state,
      names_to = "new_state",
      values_to = "Prob"
    ) |>
    filter(Prob > 0)
```

```
# A tibble: 5 x 3
  state new_state     Prob
  <chr> <chr>        <dbl>
1 000 0 000 0      0.0334
2 000 0 000 1      0.676
3 000 0 001 0      0.00563
4 000 0 010 0      0.0503
5 000 0 100 0      0.235
```

Let's contrast this with the possible transitions starting from the "010 2" state, runner on second with two outs. The most likely transitions are "3 outs" (probability 0.650), "runners on first and second with two outs" (probability 0.156), and "runner on first with 2 outs" (probability 0.074).

```
    P_matrix |>
      as_tibble(rownames = "state") |>
      filter(state == "010 2") |>
      pivot_longer(
        cols = -state,
        names_to = "new_state",
        values_to = "Prob"
      ) |>
      filter(Prob > 0)
```

```
# A tibble: 8 x 3
  state new_state     Prob
  <chr> <chr>        <dbl>
1 010 2 000 2      0.0233
2 010 2 001 2      0.00587
3 010 2 010 2      0.0576
4 010 2 011 2      0.000451
5 010 2 100 2      0.0745
6 010 2 101 2      0.0325
7 010 2 110 2      0.156
8 010 2 3         0.650
```

9.2.4   Simulating the Markov chain

One can simulate this Markov chain model a large number of times to obtain the distribution of runs scored in a half-inning of 2016 baseball. The first step is to construct a matrix giving the runs scored in all possible transitions between states. Let $N_{runners}$ denote the number of runners in a state and O denote the number of outs. Because every player who has already batted in the inning is either on base, out, or has scored, for a batting play, the number of runs scored is equal to

$$runs = (N_{runners}^{(b)} + O^{(b)} + 1) - (N_{runners}^{(a)} + O^{(a)}).$$

In other words, the runs scored is the sum of runners and outs before (b) the play minus the sum of runners and outs after (a) the play plus one. For example, suppose there are runners on first and second with one out, and after the play, there is a runner on second with two outs. The number of runs scored is equal to

$$runs = (2 + 1 + 1) - (1 + 2) = 1.$$

We define a new function `num_havent_scored()` which takes a state as input and returns the sum of the number of runners and outs. We then apply this function across all the possible states (using the `map_int()` function) and the corresponding sums are stored in the vector `runners_out`.

```
num_havent_scored <- function(s) {
  s |>
    str_split("") |>
    pluck(1) |>
    as.numeric() |>
    sum(na.rm = TRUE)
}

runners_out <- T_matrix |>
  row.names() |>
  set_names() |>
  map_int(num_havent_scored)
```

The `outer()` function with the—(subtraction) operation performs the runs calculation for all possible pairs of states and the resulting matrix is stored in the matrix `R_runs`. If one inspects the matrix `R_runs`, one will notice some negative values and some strange large positive values. But this is not a concern since the corresponding transitions, for example a movement between a "000 0" state and a "000 2" state in one batting play, are not possible. To make the matrix square, we add an additional column of zeros to this run matrix using the `cbind()` function.

```
R_runs <- outer(
  runners_out + 1,
  runners_out,
  FUN = "-"
) |>
  cbind("3" = rep(0, 24))
```

We are now ready to simulate a half-inning of baseball using a new function
`simulate_half_inning()`. The inputs are the probability transition matrix
P, the run matrix R, and the starting state s (an integer between 1 and 24).
The output is the number of runs scored in the half-inning.

```
simulate_half_inning <- function(P, R, start = 1) {
  s <- start
  path <- NULL
  runs <- 0
  while (s < 25) {
    s_new <- sample(1:25, size = 1, prob = P[s, ])
    path <- c(path, s_new)
    runs <- runs + R[s, s_new]
    s <- s_new
  }
  runs
}
```

There are two key statements in this simulation. If the current state is s, the
function `sample()` will simulate a new state using the s row in the transition
matrix P; the new state is denoted s_new. The total number of runs scored in
the inning is updated using the value in the s row and the s_new column of
the runs matrix R.

Using the `map_int()` function, one can simulate a large number of half-innings
of baseball. In the below code, we simulate 10,000 half-innings starting with
no runners and no outs (state 1), collecting the runs scored in the vector
`simulated_runs`. The `set.seed()` function sets the random number seed so
the reader can reproduce the results of this particular simulation by running
this code.

```
set.seed(111653)
simulated_runs <- 1:10000 |>
  map_int(~simulate_half_inning(T_matrix, R_runs))
```

To find the possible runs scored in a half-inning, we use the `table()` function
to tabulate the values in `simulated_runs`.

```
table(simulated_runs)
```

```
simulated_runs
   0    1    2    3    4    5    6    7    8    9
7364 1437  651  324  126   50   34   10    2    2
```

In our 10,000 simulations, five or more runs scored in $50 + 34 + 10 + 2 + 2$ = 98 half-innings, so the chance of scoring five or more runs would be 98 / 10,000 = 0.0098. This calculation can be checked using the `sum()` function.

```
sum(simulated_runs >= 5) / 10000
```

[1] 0.0098

We compute the mean number of runs scored by applying the `mean()` function to `simulated_runs`.

```
mean(simulated_runs)
```

[1] 0.477

Over the 10,000 half-innings, an average of 0.477 runs were scored.

To understand the runs potential of different runners and outs situations, one can repeat this simulation procedure for other starting states. We write a function `runs_j()` to compute the mean number of runs scored starting with state j. Using the `map_int()` function, we apply the function `runs_j()` over all of the possible starting states 1 through 24. The output is a vector of mean runs scored stored in the `mean_run_value` column. These values are displayed below as a simulated expected run matrix (see Section 5.1).

```
runs_j <- function(j) {
  1:10000 |>
    map_int(~simulate_half_inning(T_matrix, R_runs, j)) |>
    mean()
}

erm_2016_mc <- tibble(
  state = row.names(T_matrix),
  mean_run_value = map_dbl(1:24, runs_j)
) |>
  mutate(
    bases = str_sub(state, 1, 3),
    outs_ct = as.numeric(str_sub(state, 5, 5))
  ) |>
  select(-state)
```

```
erm_2016_mc |>
  pivot_wider(names_from = outs_ct, values_from = mean_run_value)
```

```
# A tibble: 8 x 4
  bases   `0`   `1`   `2`
  <chr> <dbl> <dbl> <dbl>
1 000   0.481 0.255 0.103
2 001   1.32  0.925 0.338
3 010   1.14  0.640 0.295
4 011   1.93  1.30  0.474
5 100   0.855 0.500 0.211
6 101   1.71  1.14  0.425
7 110   1.39  0.875 0.406
8 111   2.19  1.46  0.667
```

Recall that our simulation model is based only on batting plays. To understand the effect of non-batting plays (stealing, caught stealing, wild pitches, etc.) on run scoring, we compare this run expectancy matrix with the one found in Chapter 5 using all batting and non-batting plays. Their difference is the contribution of non-batting plays to the average number of runs scored.

```
erm_2016 <- read_rds(here::here("data/erm2016.rds"))
```

```
erm_2016 |>
  inner_join(erm_2016_mc, join_by(bases, outs_ct)) |>
  mutate(
    run_value_diff = round(mean_run_value.x - mean_run_value.y, 2)
  ) |>
  select(bases, outs_ct, run_value_diff) |>
  pivot_wider(names_from = outs_ct, values_from = run_value_diff)
```

```
# A tibble: 8 x 4
# Groups:   bases [8]
  bases   `0`   `1`   `2`
  <chr> <dbl> <dbl> <dbl>
1 000    0.02  0.01  0
2 001    0.03  0.01  0.03
3 010    0     0.03  0.02
4 011    0     0.06  0.07
5 100    0     0.01  0.01
6 101    0.02  0.06  0.05
7 110    0.06  0.05  0.01
8 111   -0.08  0.07  0.03
```

Note that most of the values of the difference are positive, indicating that these non-batting plays generally do create runs. We note that the largest values tend to occur in situations when there is a runner on third who can score on a wild pitch or passed ball.

9.2.5 Beyond run expectancy

By using properties of Markov chains, it is straightforward to use the transition matrix to learn more about the movement through the runners/outs states.

By multiplying the probability matrix `P_matrix` by itself three times, we can learn about the likelihood of the state of the inning after three plate appearances. In R, *matrix multiplication* is indicated by the `%*%` symbol. The result is stored in the matrix `P_matrix_3`.

```
P_matrix_3 <- P_matrix %*% P_matrix %*% P_matrix
```

The first row of `P_matrix_3` gives the probabilities of being in each of the 25 states after three hitters starting at the "000 0" state. We round these values to three decimal places, sort from largest to smallest, and display the largest values.

```
P_sorted <- P_matrix_3 |>
  as_tibble(rownames = "state") |>
  filter(state == "000 0") |>
  pivot_longer(
    cols = -state, names_to = "new_state", values_to = "Prob"
  ) |>
  arrange(desc(Prob))
P_sorted |>
  slice_head(n = 6)
```

```
# A tibble: 6 x 3
  state new_state   Prob
  <chr> <chr>       <dbl>
1 000 0 3          0.372
2 000 0 100 2      0.241
3 000 0 110 1      0.0815
4 000 0 010 2      0.0739
5 000 0 000 2      0.0529
6 000 0 001 2      0.0286
```

After three PAs, the most likely outcomes are three outs (probability 0.372), runner on first with 2 outs (probability 0.241), and runners on first and second with one out (probability 0.081).

It is also easy to learn about the number of visits to all runner-outs states. Define the matrix Q to be the 24-by-24 submatrix found from the transition matrix by removing the last row and column (the three outs state). By subtracting the matrix Q from the identity matrix and taking the inverse of the result, we obtain the fundamental matrix N of an absorbing Markov chain. (The `diag()` function is used to construct the identity matrix and the function `solve()` takes the matrix inverse.)

```
Q <- P_matrix[-25, -25]
N <- solve(diag(rep(1, 24)) - Q)
```

To understand the fundamental matrix, we display the beginning entries of the first row of the matrix.

```
N_0000 <- round(N["000 0", ], 2)
head(N_0000, n = 6)
```

```
000 0 000 1 000 2 001 0 001 1 001 2
 1.05  0.75  0.60  0.01  0.03  0.05
```

Starting at the beginning of the inning (the "000 0" state), the average number of times the inning will be in the "000 0" state is 1.05, the average number of times in the "000 1" state is 0.75, the average number of times in the "000 2" state is 0.6, and so on. By using the `sum()` function, we find the average number of states that are visited.

```
sum(N_0000)
```

```
[1] 4.27
```

In other words, the average number of plate appearances in a half-inning (before three outs) is 4.27.

We can compute the average number of batting plays until three outs for all starting states by multiplying the fundamental matrix N by a column vector of ones. The vector of average number of plays is stored in the variable `avg_num_plays` and eight values of this vector are displayed.

```
avg_num_plays <- N %*% rep(1, 24) |>
  t() |>
  round(2)
avg_num_plays[,1:8]
```

```
000 0 000 1 000 2 001 0 001 1 001 2 010 0 010 1
 4.27  2.87  1.46  4.33  2.99  1.53  4.34  2.93
```

This tells us the length of the remainder of the inning, on average, starting with each possible state. For example, starting at the bases empty, one out state, we expect on average to have 2.87 more batters. In contrast, with a runner on third with two outs, we expect to have 1.53 more batters.

9.2.6   Transition probabilities for individual teams

The transition probability matrix describes movements between states for an average team. Certainly, these probabilities will vary for teams of different batting abilities, and the probabilities will also vary against teams of different pitching abilities. We focus on different batting teams and discuss how to obtain good estimates of the transition probabilities for all teams.

To get the relevant data, a new variable `batting_team` needs to be defined that gives the batting team in each half-inning. By use of the `str_sub()` function, we define the home team variable `home_team_id`, and an `if_else()` function is used to define the batting team.

```
retro2016_complete <- retro2016_complete |>
  mutate(
    home_team_id = str_sub(game_id, 1, 3),
    batting_team = if_else(
      bat_home_id == 0,
      away_team_id,
      home_team_id
    )
  )
```

By use of the `group_by()` and `count()` functions, we construct a data frame `T_team` giving the counts of each team in the transitions from the current to new states.

```
T_team <- retro2016_complete |>
  group_by(batting_team, state, new_state) |>
  count()
```

For example, the filtering for `batting_team` equal to `ANA` gives the transition counts for Anaheim in the 2016 season.

```
T_team |>
  filter(batting_team == "ANA") |>
  slice_head(n = 6)
```

```
# A tibble: 192 x 4
# Groups:   batting_team, state, new_state [192]
  batting_team state new_state      n
```

	<chr>	<chr> <chr>	<int>
1	ANA	000 0 000 0	40
2	ANA	000 0 000 1	1007
3	ANA	000 0 001 0	9
4	ANA	000 0 010 0	75
5	ANA	000 0 100 0	359
6	ANA	000 1 000 1	31
7	ANA	000 1 000 2	720
8	ANA	000 1 001 1	3
9	ANA	000 1 010 1	54
10	ANA	000 1 100 1	261

```
# i 182 more rows
```

If one is interested in comparing run productions for different batting teams, it is necessary to make some adjustments to the team transition probability matrices to get realistic predictions of performance. To illustrate the problem, we focus on transitions from the "100 2" state. We store the transition counts in the data frame T_team_S using the `tally()` function and display a few rows of this table below for six of the teams.

```
T_team_S <- retro2016_complete |>
  filter(state == "100 2") |>
  group_by(batting_team, state, new_state) |>
  tally()

T_team_S |>
  ungroup() |>
  sample_n(size = 6)
```

```
# A tibble: 6 x 4
  batting_team state new_state     n
  <chr>        <chr> <chr>     <int>
1 MIN          100 2 010 2        15
2 CIN          100 2 101 2        19
3 NYN          100 2 101 2        14
4 SEA          100 2 010 2         6
5 DET          100 2 011 2         9
6 DET          100 2 101 2        15
```

For some of the less common transitions, there is much variability in the counts across teams and this causes the corresponding team transition probabilities to be unreliable. If p^{TEAM} represents the team's transition probabilities for a particular team, and p^{ALL} are the average transition probabilities, then a better estimate at the team's probabilities has the form

$$p^{EST} = \frac{n}{n+K} p^{TEAM} + \frac{K}{n+K} p^{ALL},$$

where n is the number of transitions for the team and K is a smoothing count. The description of the methodology is beyond the scope of this book, but in this case a smoothing count of $K = 1274$ leads to a good estimate at the team's transition probabilities. (The choice of K depends on the starting state.)

This method is illustrated for Washington's transition counts starting from the "100 2" state. In the data frame T_WAS, the transition counts are stored in the variable n, and the corresponding proportions are stored in p. Similarly, for all teams, n and p are the counts and proportions in the data frame T_all.

```
T_WAS <- T_team_S |>
  filter(batting_team == "WAS") |>
  mutate(p = n / sum(n))

T_all <- retro2016_complete |>
  filter(state == "100 2") |>
  group_by(new_state) |>
  tally() |>
  mutate(p = n / sum(n))
```

We compute the improved estimate at Washington's transition proportions using the formula and store the results in p_EST. The three sets of proportions (Washington, overall, and improved) are displayed in a data frame.

```
T_WAS |>
  inner_join(T_all, by = "new_state") |>
  mutate(
    p_EST = (n.x / (1274 + n.x)) * p.x + (1274 / (1274 + n.x)) * p.y
  ) |>
  select(batting_team, new_state, p.x, p.y, p_EST)
```

```
# A tibble: 8 x 6
# Groups:   batting_team, state [1]
  state batting_team new_state      p.x       p.y     p_EST
  <chr> <chr>        <chr>        <dbl>     <dbl>     <dbl>
1 100 2 WAS          000 2      0.0319    0.0291    0.0291
2 100 2 WAS          001 2      0.00532   0.00577   0.00577
3 100 2 WAS          010 2      0.0213    0.0220    0.0219
4 100 2 WAS          011 2      0.0213    0.0220    0.0219
5 100 2 WAS          100 2      0.00266   0.000775  0.000776
6 100 2 WAS          101 2      0.0452    0.0435    0.0435
7 100 2 WAS          110 2      0.184     0.195     0.194
8 100 2 WAS          3          0.689     0.682     0.683
```

Note that the improved transition proportions are a compromise between the team's proportions and the overall values. For example, for a transition from the state "100 2" to "010 2", the Washington value is 0.0213, the overall value is 0.0220, and the improved value 0.0219 falls between the Washington and overall values. This method is especially helpful for particular transitions such as "100 2" to "100 2", which may not occur for one team in this season but for which we know there is a positive chance of these transitions happening in the future.

This smoothing method can be applied for all teams and all rows of the transition matrix to obtain improved estimates of teams' probability transition matrices. With the team transition matrices computed in this way, one can explore the run-scoring behavior of individual batting teams.

9.3 Simulating a Baseball Season

9.3.1 The Bradley-Terry model

An attractive method of modeling paired comparison data such as baseball games is the Bradley-Terry model. We illustrate this modeling technique via simulation for the 1968 Major League Baseball season when the regular season and playoff system had a relatively simple structure. It is straightforward to adapt these methods to the present baseball season with a more complicated schedule and playoff system.

In 1968, there were 20 teams, 10 in the National League and 10 in the American League. Suppose each team has a talent or ability to win a game. The talents for the 20 teams are represented by the values $T_1, ..., T_{20}$. We assume that the talents are distributed from a normal curve model with mean 0 and standard deviation s_T. A team of average ability would have a talent value close to zero, "good" teams would have positive talents, and bad teams would have negative talents. Suppose team A plays team B in a single game. By the Bradley-Terry model, the probability team A wins the game is given by the logistic function

$$P(A\,wins) = \frac{\exp(T_A)}{\exp(T_A) + \exp(T_B)}.$$

This model is closely related to the log5 method developed by Bill James in his *Baseball Abstract* books in the 1980s (see, for example, James (1982)). If P_A and P_B are the winning percentages of teams A and B, then James' formula is given by

$$P(A\,wins) = \frac{P_A/(1 - P_A)}{P_A/(1 - P_A) + P_B/(1 - P_B)}.$$

Comparing the two formulas, one sees that the log5 method is a special case of the Bradley-Terry model where a team's talent T is set equal to the log odds of winning $\log(P/(1-P))$. A team with a talent $T = 0$ will win (in the long run) half of its games ($P = 0.5$). In contrast, a team with talent $T = 0.2$ will win (using the log 5 values) approximately 55% of its games and a team with talent $T = -0.2$ will win 45% of its games.

Using this model, one can simulate a baseball season as follows.

1. Construct the 1968 baseball schedule. In this season, each of the 10 teams in each league play each other team in the same league 18 games, where 9 games are played in each team's ballpark. (There was no interleague play in 1968.)
2. Simulate 20 talents from a normal distribution with mean 0 and standard deviation s_T. The value of s_T is chosen so that the simulated season winning percentages from this model resemble the actual winning percentages during this season.
3. Using the probability formula and the talent values, one computes the probabilities that the home team wins all games. By a series of coin flips with these probabilities, one determines the winners of all games.
4. Determine the winner of each league (ties need to be broken by some random mechanism) and play a best-of-seven World Series using winning probabilities computed using the Bradley-Terry model and the two talent numbers.

9.3.2 Making up a schedule

The first step in the simulation is to construct the schedule of games. We wrote a short function `make_schedule()` to help with this task. The inputs are the vector of team names `teams` and the number of games `k` that will be played between two teams in the first team's home park. The output is a data frame where each row corresponds to a game and `Home` and `Visitor` give the names of the home and visiting teams. The `rep()` function, which generates repeated copies of a vector, is used several times in this function.

```
make_schedule <- function(teams, k) {
  num_teams <- length(teams)
  Home <- rep(rep(teams, each = num_teams), k)
  Visitor <- rep(rep(teams, num_teams), k)
  tibble(Home = Home, Visitor = Visitor) |>
    filter(Home != Visitor)
}
```

This function is used to construct the schedule for the 1968 season. Two vectors `NL` and `AL` are constructed containing abbreviations for the National League and American League teams. We apply the function `make_schedule()` twice, once for each league, using `k = 9` since one team hosts another team nine games. We use the `list_rbind()` function to paste together the NL and AL schedules, creating the data frame `schedule`.

```
library(Lahman)
teams_68 <- Teams |>
  filter(yearID == 1968) |>
  select(teamID, lgID) |>
  mutate(teamID = as.character(teamID)) |>
  group_by(lgID)

schedule <- teams_68 |>
  group_split() |>
  set_names(pull(group_keys(teams_68), "lgID")) |>
  map(~make_schedule(teams = .x$teamID, k = 9)) |>
  list_rbind(names_to = "lgID")
dim(schedule)
```

```
[1] 1620    3
```

Note that `schedule` has $\frac{162 \cdot 20}{2}$ rows, since each game involves two teams.

9.3.3 Simulating talents and computing win probabilities

The next step is to compute the win probabilities for all of the games in the season schedule. The team talents are assumed to come from a normal distribution with mean 0 and standard deviation `s_talent`, which we assign `s_talent = 0.20`. (Recall that this value of the standard deviation is chosen so that the season team win percentages generated from the model resemble the actual team win percentages.) We simulate the talents using the function `rnorm()` that assigns the talents to the 20 teams. By use of two applications of the `inner_join()` function, we add the team talents to the schedule data frame; the new data frame is called `schedule_talent`.

```
s_talent <- 0.20
teams_68 <- teams_68 |>
  mutate(talent = rnorm(10, 0, s_talent))

schedule_talent <- schedule |>
  inner_join(teams_68, join_by(lgID, Home == teamID)) |>
  rename(talent_home = talent) |>
  inner_join(teams_68, join_by(lgID, Visitor == teamID)) |>
  rename(talent_visitor = talent)
```

Last, once we have the talents for the home and visiting teams for all games, we apply the Bradley-Terry model to compute home team winning probabilities for all games; these probabilities are stored in the variable `prob_home`.

```
schedule_talent <- schedule_talent |>
  mutate(
    prob_home = exp(talent_home) /
      (exp(talent_home) + exp(talent_visitor))
  )
```

The first six rows of the data frame `schedule_talent` are displayed below, where one sees the games scheduled, the talents of the home and away teams, and the probability that the home team wins the matchup.

```
slice_head(schedule_talent, n = 6)
```

```
# A tibble: 6 x 6
  lgID  Home  Visitor talent_home talent_visitor prob_home
  <chr> <chr> <chr>         <dbl>          <dbl>     <dbl>
1 AL    BAL   BOS           0.197          0.269     0.482
2 AL    BAL   CAL           0.197         -0.230     0.605
3 AL    BAL   CHA           0.197         -0.00924   0.551
4 AL    BAL   CLE           0.197         -0.185     0.594
5 AL    BAL   DET           0.197          0.409     0.447
6 AL    BAL   MIN           0.197         -0.208     0.600
```

9.3.4   Simulating the regular season

To simulate an entire season of games, we perform a series of coin flips, where the probability the home team wins depends on the winning probability. The function `rbinom()` performs the coin flips for the 1620 scheduled games; the outcomes are a sequence of 0s and 1s. By use of the `if_else()` function, we define the `winner` variable to be the `Home` team if the outcome is 1, and the `Visitor` otherwise.

```
schedule_talent <- schedule_talent |>
  mutate(
    outcome = rbinom(nrow(schedule_talent), 1, prob_home),
    winner = if_else(outcome == 1, Home, Visitor)
  )
```

The teams, home win probabilities, and outcomes of the first six games are displayed below.

```
schedule_talent |>
  select(Visitor, Home, prob_home, outcome, winner) |>
  slice_head(n = 6)
```

```
# A tibble: 6 x 5
  Visitor Home  prob_home outcome winner
  <chr>   <chr>     <dbl>   <int> <chr>
1 BOS     BAL       0.482       0 BOS
2 CAL     BAL       0.605       0 CAL
3 CHA     BAL       0.551       1 BAL
4 CLE     BAL       0.594       1 BAL
5 DET     BAL       0.447       0 DET
6 MIN     BAL       0.600       0 MIN
```

How did the teams perform during this particular simulated season? Using the `group_by()` and `summarize()` functions, we find the number of wins for all teams. We collect this information together with the team names in the data frame WIN, and use the `inner_join()` function to combine the season results with the team talents to create the data frame `results`.

```
results <- schedule_talent |>
  group_by(winner) |>
  summarize(Wins = n()) |>
  inner_join(teams_68, by = c("winner" = "teamID"))
```

9.3.5 Simulating the post-season

After the regular season, one can simulate the post-season series. We write a function `win_league()` that simulates a league championship. The inputs are the data frame `res` of teams and win totals. By use of the `min_rank()` function, we identify the teams that have the largest number of wins in each league. If one team has the maximum number, then an indicator variable `is_winner_lg` is created, which is 1 for that particular team. In order to avoid a tie in win totals for two or more teams, we randomly add a random `tiebreaker` quantity (that is less than 1) to every teams win total using the `runif()` function.

```
win_league <- function(res) {
  res |>
    group_by(lgID) |>
    mutate(
      tiebreaker = runif(n = length(talent)),
      wins_total = Wins + tiebreaker,
      rank = min_rank(desc(wins_total)),
      is_winner_lg = wins_total == max(wins_total)
    )
}
```

To simulate the post-season, we populate a new variable `is_winner_ws`; this is an indicator for the World Series winner. By an application of `win_league()`, we find the winners of each league. We simulate the World Series by flipping a coin seven times (`rmultinom()`), where the win probabilities are proportional to exp(*talent*). The `is_winner_ws` indicates the team winning a majority of the games.

```
sim_one <- win_league(results)

ws_winner <- sim_one |>
  filter(is_winner_lg) |>
  ungroup() |>
  mutate(
    outcome = as.numeric(rmultinom(1, 7, exp(talent))),
    is_winner_ws = outcome > 3
  ) |>
  filter(is_winner_ws) |>
  select(winner, is_winner_ws)

sim_one |>
  left_join(ws_winner, by = c("winner")) |>
  replace_na(list(is_winner_ws = 0))
```

```
# A tibble: 20 x 9
# Groups:   lgID [2]
   winner  Wins lgID     talent tiebreaker wins_total  rank
   <chr>  <int> <fct>     <dbl>      <dbl>      <dbl> <int>
 1 ATL       83 NL      -0.215      0.867        83.9     5
 2 BAL       86 AL       0.197      0.0260       86.0     3
 3 BOS       99 AL       0.269      0.354        99.4     2
 4 CAL       62 AL      -0.230      0.936        62.9    10
 5 CHA       85 AL      -0.00924    0.246        85.2     4
 6 CHN       85 NL      -0.107      0.829        85.8     4
 7 CIN       81 NL      -0.0612     0.264        81.3     6
 8 CLE       71 AL      -0.185      0.290        71.3     9
 9 DET      100 AL       0.409      0.0841      100.      1
10 HOU       90 NL      -0.0871     0.845        90.8     2
11 LAN       63 NL      -0.326      0.729        63.7    10
12 MIN       74 AL      -0.208      0.667        74.7     7
13 NYA       78 AL       0.0424     0.422        78.4     6
14 NYN       80 NL       0.100      0.569        80.6     7
15 OAK       82 AL       0.287      0.927        82.9     5
16 PHI       93 NL       0.265      0.819        93.8     1
17 PIT       71 NL      -0.146      0.0460       71.0     9
18 SFN       88 NL       0.249      0.119        88.1     3
19 SLN       76 NL      -0.348      0.425        76.4     8
20 WS2       73 AL       0.0842     0.730        73.7     8
# i 2 more variables: is_winner_lg <lgl>, is_winner_ws <lgl>
```

9.3.6 Function to simulate one season

It is convenient to place all of these commands including the functions `make_schedule()` and `win_league()` in a single function `one_simulation_68()`, which you can find in the **abdwr3edata** package. The only input is the standard deviation `s_talent` that describes the spread of the normal talent distribution. The output is a data frame containing the teams, talents, number of season wins, and success in the post-season. We illustrate simulating one season and display the data frame `results_1` that is returned.

```
library(abdwr3edata)
set.seed(111653)
results_1 <- one_simulation_68(0.20)
results_1
```

```
# A tibble: 20 x 6
    Team  Wins League   Talent Winner.Lg Winner.WS
    <chr> <int>  <dbl>    <dbl>     <dbl>     <dbl>
 1 SFN      93      1 -0.0591         1         0
 2 PHI      93      1 -0.00979        0         0
 3 LAN      87      1  0.00406        0         0
 4 HOU      84      1 -0.117          0         0
 5 SLN      80      1 -0.128          0         0
 6 ATL      79      1 -0.100          0         0
 7 CIN      79      1 -0.235          0         0
 8 NYN      76      1 -0.269          0         0
 9 CHN      76      1 -0.0199         0         0
10 PIT      63      1 -0.313          0         0
11 NYA     100      2  0.284          1         1
12 DET      93      2  0.379          0         0
13 CHA      87      2  0.139          0         0
14 BOS      86      2 -0.102          0         0
15 WS2      84      2  0.0915         0         0
16 OAK      82      2 -0.0622         0         0
17 CAL      78      2 -0.129          0         0
18 BAL      74      2 -0.0728         0         0
19 MIN      65      2 -0.207          0         0
20 CLE      61      2 -0.292          0         0
```

We write a new function `display_standings()` to put the season wins in a more familiar standings format. The inputs to this function are the `results_1` data frame and the league indicator.

```
display_standings <- function(data, league) {
  data |>
    filter(League == league) |>
    select(Team, Wins) |>
    mutate(Losses = 162 - Wins) |>
    arrange(desc(Wins))
}
```

We then apply this function twice (once for each league) using `map()` and then use the `bind_cols()` function to combine the two standings into a single data frame. The league champions and the World Series winner are also displayed below.

```
map(1:2, display_standings, data = results_1) |>
  bind_cols()
```

```
# A tibble: 10 x 6
   Team...1 Wins...2 Losses...3 Team...4 Wins...5 Losses...6
   <chr>       <int>      <dbl> <chr>       <int>      <dbl>
 1 SFN            93         69 NYA           100         62
 2 PHI            93         69 DET            93         69
 3 LAN            87         75 CHA            87         75
 4 HOU            84         78 BOS            86         76
 5 SLN            80         82 WS2            84         78
 6 ATL            79         83 OAK            82         80
 7 CIN            79         83 CAL            78         84
 8 NYN            76         86 BAL            74         88
 9 CHN            76         86 MIN            65         97
10 PIT            63         99 CLE            61        101
```

```
results_1 |>
  filter(Winner.Lg == 1) |>
  select(Team, Winner.WS)
```

```
# A tibble: 2 x 2
  Team  Winner.WS
  <chr>     <dbl>
1 SFN           0
2 NYA           1
```

In this particular simulated season, the Philadelphia Phillies (PHI) and the San Francisco Giants (SFN) tied for the National League title with 93 wins and the New York Yankees (NYA) won the American League with 100 wins. The Yankees defeated the Giants in the World Series. The team with the best talent in this season was Detroit (talent equal to 0.379) and they lost in the

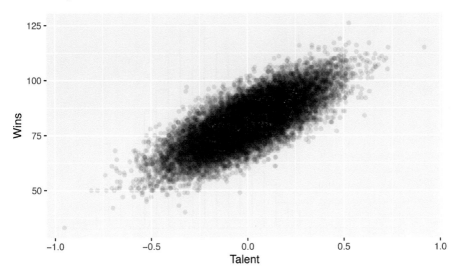

FIGURE 9.1
Scatterplot of team talents and wins for many season simulations.

ALCS. In other words the "best team in baseball" was not the most successful during this simulated season. We will shortly see if the best team typically wins the World Series.

9.3.7 Simulating many seasons

One can learn about the relationship between a team's ability and its season performance by simulating many seasons of baseball. To simulate 1000 seasons, we use the `rep()` function to create a vector of length 1000, then use `map()` to repeatedly apply the `one_simulation_68()` function to this vector, storing the output in `many_results`.

```
set.seed(111653)
many_results <- rep(0.20, 1000) |>
  map(one_simulation_68) |>
  list_rbind()
```

The data frame `many_results` contains the talent number and number of wins for $1000 \times 20 = 20{,}000$ teams. By use of the `geom_point()` function using the `alpha = 0.05` argument, we construct a "smoothed" scatterplot of `Talent` and `Wins` in Figure 9.1.

```
ggplot(many_results, aes(Talent, Wins)) +
  geom_point(alpha = 0.05)
```

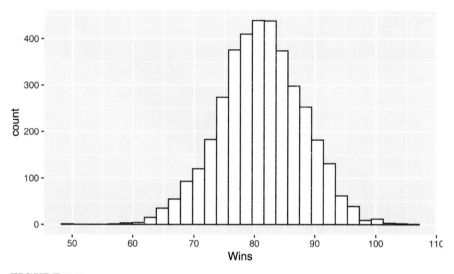

FIGURE 9.2
Histogram of number of wins of teams of average talents in the simulations.

As expected, there is a positive trend in the graph, indicating that better teams tend to win more games. But there is much vertical spread in the scatterplot, which says that the relationship between talent and wins is not strong.

To reinforce the last point, suppose we focus on "average" teams that have a talent number between -0.05 and 0.05. Using the `filter()` function, we isolate the talent and wins data for these average teams. A histogram of the season wins for these teams is shown in Figure 9.2.

```
many_results |>
    filter(Talent > -0.05, Talent < 0.05) |>
    ggplot(aes(Wins)) +
    geom_histogram(color = crcblue, fill = "white")
```

One expects these average teams to win about 81 games. But what is surprising is the variability in the win totals—average teams can regularly have win totals between 70 and 90, and it is possible (but not likely) to have a win total close to 100.

What is the relationship between a team's talent and its post-season success? Consider first the relationship between a team's talent (variable `Talent`) and winning the league (the variable `Winner.Lg`). Since `Winner.Lg` is a binary (0 or 1) variable, a common approach for representing this relationship is a logistic model; this is a generalization of the usual regression model where the response variable is binary instead of continuous. We use the `glm()` function with the `family` argument set to `binomial` to fit a logistic model; the output is stored in the variable `fit1`. In a similar fashion, we use a logistic model to model the relationship between winning the World Series (variable `Winner.WS`) and the talent; the output is in the variable `fit2`.

```
fit1 <- glm(
  Winner.Lg ~ Talent,
  data = many_results, family = binomial
)
fit2 <- glm(
  Winner.WS ~ Talent,
  data = many_results, family = binomial
)
```

A logistic model has the form

$$p = \frac{\exp(a + bT)}{1 + \exp(a + bT)},$$

where T is a team's talent, (a, b) are the regression coefficients, and p is the probability of the event.

In the following code, we generate a vector of plausible values of a team's talent and store them in the vector `talent_values`. We then compute fitted probabilities of winning the pennant and winning the World Series using the `predict()` function. (The `type = "response"` argument will map values of $a + bT$ to the probability scale.) Then we construct a graph of talent against probability using the `geom_line()` function where the color of the line corresponds to the type of achievement. The completed graph is displayed in Figure 9.3.

```
tdf <- tibble(
  Talent = seq(-0.4, 0.4, length.out = 100)
)
tdf |>
  mutate(
    Pennant = predict(fit1, newdata = tdf, type = "response"),
    `World Series` = predict(fit2, newdata = tdf, type = "response")
  ) |>
  pivot_longer(
    cols = -Talent,
    names_to = "Outcome",
    values_to = "Probability"
  ) |>
  ggplot(aes(Talent, Probability, color = Outcome)) +
  geom_line() + ylim(0, 1) +
  scale_color_manual(values = crc_fc)
```

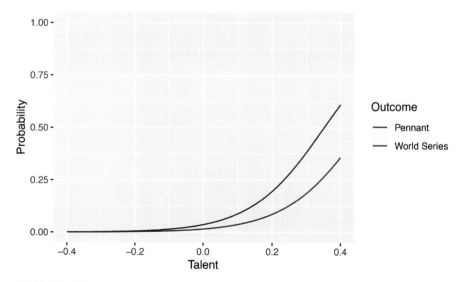

FIGURE 9.3
Probability of winning the pennant and winning the World Series for teams of different talents.

As expected, the chance of a team winning the pennant (solid line) increases as a function of the talent. An average team with $T = 0$ has only a small chance of winning the pennant; an excellent team with a talent close to 0.4 has about a 60% chance of winning the pennant. The probabilities of winning the World Series (represented by a dashed line) are substantially smaller than the chances of winning the pennant. For example, this excellent ($T = 0.4$) team has only about a 35% chance of winning the World Series. In fact, it can be demonstrated that the team winning the World Series is likely not to be the team with the best talent (largest value of T).

9.4 Further Reading

A general description of the Markov chain probability model is contained in Kemeny and Snell (1960). Pankin (1987) and Bukiet, Harold, and Palacios (1997) illustrate the use of Markov chains to model baseball. Chapter 9 of Albert (2017) gives an introductory description of Markov chains and illustrates the construction and use of the transition matrix using 1987 season data. The Bradley-Terry model (Bradley and Terry 1952) is a popular statistical model for paired comparisons. Chapter 9 of Albert and Bennett (2003) describes the application of the Bradley-Terry model for baseball team competition. The use of R in simulation is introduced in Chapter 11 of Albert and Rizzo (2012). Lopez, Matthews, and Baumer (2018) use a Bradley-Terry state space model to address the question of how often the best teams win.

9.5 Exercises

1. A Simple Markov Chain

Suppose one is interested only in the number of outs in an inning. There are four possible states in an inning (0 outs, 1 out, 2 outs, and 3 outs) and you move between these states in each plate appearance. Suppose at each PA, the chance of not increasing the number of outs is 0.3, and the probability of increasing the outs by one is 0.7. The following R code puts the transition probabilities of this Markov chain in a matrix P.

```
P <- matrix(c(.3, .7,  0,  0,
               0, .3, .7,  0,
               0,  0, .3, .7,
               0,  0,  0,  1), 4, 4, byrow = TRUE)
```

a. If one multiplies the matrix P by itself P to obtain the matrix P2:

```
P2 <- P %*% P
```

The first row of P2 gives the probabilities of moving from 0 outs to each of the four states after two plate appearances. Compute P2. Based on this computation, find the probability of moving from 0 outs to 1 out after two plate appearances.

b. The fundamental matrix N is computed as

```
N <- solve(diag(c(1, 1, 1)) - P[-4, -4])
```

The first row gives the average number of PAs at 0 out, 1 out, and 2 outs in an inning. Compute N and find the average number of PAs in one inning in this model.

2. A Simple Markov Chain, Continued

The following function `simulate_half_inning()` will simulate the number of plate appearances in a single half-inning of the Markov chain model described in Exercise 1 where the input P is the transition probability matrix.

```
simulate_half_inning <- function(P) {
  s <- 1
  path <- NULL
  while(s < 4){
    s_new <- sample(1:4, 1, prob = P[s, ])
    path <- c(path, s_new)
    s <- s_new
  }
  length(path)
}
```

a. Use the `map()` function to simulate 1000 half-innings of this Markov chain and store the lengths of these simulated innings in the vector `lengths`.

b. Using this simulated output, find the probability that a half-inning contains exactly four plate appearances.

c. Use the simulated output to find the average number of PAs in a half-inning. Compare your answer with the exact answer in Exercise 1, part (b).

3. Simulating a Half Inning

In Section 9.2.4, the expected number of runs as calculated for each one of the 24 possible runners-outs situations using data from the 2016 season. To see how these values can change across seasons, download play-by-play data from Retrosheet for the 1968 season, construct the probability transition matrix, simulate 10,000 half-innings from each of the 24 situations, and compute the run expectancy matrix. Compare this 1968 run expectancy matrix with the one computed using 2016 data.

4. Simulating the 1950 Season

Suppose you are interested in simulating the 1950 regular season for the National League. In this season, the team abbreviations were "PHI", "BRO", "NYG", "BSN", "STL", "CIN", "CHC", and "PIT" and each team played every other team 22 games (11 games at each park).

a. Using the function `make_schedule()`, construct the schedule of games for this NL season.

b. Suppose the team talents follow a normal distribution with mean 0 and standard deviation 0.25. Using the Bradley-Terry model, assign home win probabilities for all games on the schedule.

c. Use the `rbinom()` function to simulate the outcomes of all 616 games of the NL 1950 season.

d. Compute the number of season wins for all teams in your simulation.

5. Simulating the 1950 Season, Continued

a. Write a function to perform the simulation scheme described in Exercise 4. Have the function return the team with the largest talent and the team with the most wins. (If there is a tie for the league pennant, have the function return one of the best teams at random.)

b. Repeat this simulation for 1000 seasons, collecting the most talented team and the most successful team for all seasons.

c. Based on your simulations, what is the chance that the most talented team wins the pennant?

6. Simulating the World Series

a. Write a function to simulate a World Series. The input is the probability p that the AL team will defeat the NL team in a single game.

b. Suppose an AL team with talent 0.40 plays a NL team with talent 0.25. Using the Bradley-Terry model, determine the probability p that the AL wins a game.

c. Using the value of p determined in part (b), simulate 1000 World Series and find the probability the AL team wins the World Series.

d. Repeat parts (b) and (c) for AL and NL teams who have the same talents.

10

Exploring Streaky Performances

10.1 Introduction

Some of the most interesting phenomena in baseball are streaky or hot/cold performances by hitters and pitchers. During particular periods in the season, a particular player will hit for a high batting average, and in other periods, the player will be in a "cold streak" and all batted balls appear to be fielded for outs. In this chapter, we'll use R to explore streaky hitting performances.

One of the great hitting accomplishments in baseball history is Joe DiMaggio's 56-game hitting streak, and Section 10.2 explores DiMaggio's game-to-game hitting for the 1941 season. We use an R function to find all of DiMaggio's hitting streaks, and a moving average function to explore DiMaggio's batting average over short time intervals. Retrosheet play-by-play data records batters' performances in all plate appearances and we use this data in Section 10.3 to explore hitting streaks in individual at-bats. Suppose a hitter is going through an "0 for 20" hitting slump; should we be surprised? One way of answering this question is to find the longest hitting slumps for all hitters in a particular baseball season. A second way to understand the size of this hitting slump is to contrast this hitting with patterns of slumps under a random model. We describe a method for simulating a random pattern of hits and outs and use this method to assess if a particular player exhibits more streakiness in his hitting sequence than what one would expect by chance.

This discussion of streakiness focuses on patterns of hits and outs, and certainly the quality of an at-bat depends on more than just getting a hit. Section 10.4 discusses patterns of streakiness using the players' launch speeds among the batted balls. We look at players' mean launch speeds over groups of five games during a season. A way to describe streaky hitting behavior is to look at the variability of the five-game mean launch speed values. Using this measure of streakiness, we identify the streaky hitters during the 2016 season.

10.2   The Great Streak

10.2.1   Finding game hitting streaks

Whenever there is a discussion of great streaky performances in baseball, one has to talk about the "Great Streak" where Joe DiMaggio got a hit in 56 consecutive games during the 1941 season. Many people think that this particular hitting accomplishment is one of the few baseball records that will not be broken in our lifetimes. We use DiMaggio's game-to-game hitting data to motivate how we can use R to explore streaky performances.

In contrast with previous versions of this book, play-by-play hitting records are now available from Retrosheet for the 1941 season. You can download the entire season's worth of data using the `retrosheet_data()` function from the **baseballr** package, following the procedure outlined in Section A.1.3.

Nevertheless, Baseball-Reference gives a game-to-game hitting log for DiMaggio for this season that will suffice for our purposes. We've placed a copy of the data that contains this hitting log as the `dimaggio_1941` data object in the **abdwr3edata** package after copying-and-pasting it from the appropriate table (at the time of this writing, the fifth one on the page) on Baseball-Reference.com's website. We create a new data frame `joe`.

```
library(abdwr3edata)
joe <- dimaggio_1941
```

For each game during the season, the data frame records `AB`, the number of at-bats, and `H`, the number of hits. As a quick check that the data has been entered correctly, we compute DiMaggio's season batting average by summing the game hit totals and dividing by the total at-bats.

```
joe |> summarize(AVG = sum(H) / sum(AB))
```

```
# A tibble: 1 x 1
    AVG
  <dbl>
1 0.357
```

The result agrees with DiMaggio's published 1941 batting average of .357. (Actually, although this was a high average, it was overshadowed by Ted Williams' .406 average during the 1941 season.

A hitting streak is commonly defined as the number of consecutive games in which a player gets at least one base hit. Suppose we're interested in computing all of DiMaggio's hitting streaks for the 1941 season. Toward this goal, using the `if_else()` function, we create a new variable `had_hit` for each game that is either 1 or 0 depending on whether DiMaggio recorded at least one hit in the game.

```
joe <- joe |>
  mutate(had_hit = if_else(H > 0, 1, 0))
```

We display the values of `had_hit` that visually show DiMaggio's streaky hitting performance using the `pull()` function.

```
pull(joe, had_hit)
```

```
  [1] 1 1 1 1 1 1 1 1 0 0 0 1 1 1 0 1 1 0 0 1 0 0 0 1 1 1 0 0 1
 [30] 1 1 1 1 1 1 1 1 1 1 1 1 1 1 1 1 1 1 1 1 1 1 1 1 1 1 1 1 1
 [59] 1 1 1 1 1 1 1 1 1 1 1 1 1 1 1 1 1 1 1 1 1 1 1 1 1 1 0 1 1
 [88] 1 1 1 1 1 1 1 1 1 1 1 1 0 0 1 1 1 1 0 0 1 1 0 0 0 1 1
[117] 1 1 0 0 0 1 1 1 1 1 1 0 1 0 1 1 1 1 1 0 1 1
```

We see that DiMaggio started the season with an eight-game hitting streak, then had three games with no hits, a hitting streak of three games, and so on.

Suppose we wish to compute all hitting streaks for a particular player. This is conveniently done using the following user-defined function `streaks()`. The input to this function is a vector `y` of 0s and 1s corresponding to game results where the player was hitless (0) or received at least one hit (1). The output will be a data frame containing the lengths of all hitting streaks and all hitting slumps where the variable `values` indicates if the run is a streak or a slump. The function `rle()` from the **base** package computes the lengths and values of streaks of equal values in an input vector. We use the `as_tibble()` function to return a `tibble`.

```
streaks <- function(y) {
  x <- rle(y)
  class(x) <- "list"
  as_tibble(x)
}
```

Next, we apply this function to DiMaggio's game hit/no-hit sequence stored in the variable `had_hit`. Note that we use the `filter()` function is used to select the lengths of the hitting streaks.

```
joe |>
  pull(had_hit) |>
  streaks() |>
  filter(values == 1) |>
```

```
pull(lengths)
```

```
[1]  8  3  2  1  3 56 16  4  2  4  7  1  5  2
```

This function picks up DiMaggio's famous 56-game hitting streak. It is remarkable to note that Joe followed his 56-game streak immediately with a 16-game hitting streak.

The media is also fascinated with streaks of no-hit games. One can find DiMaggio's streaks of hitless games by using the `streaks()` function with the specification `values == 0` to select the lengths of the hitting slumps.

```
joe |>
  pull(had_hit) |>
  streaks() |>
  filter(values == 0) |>
  pull(lengths)
```

```
[1] 3 1 2 3 2 1 2 2 3 3 1 1 1
```

It is interesting that the length of the longest streak of no-hit games was only three for DiMaggio's 1941 season.

10.2.2 Moving batting averages

An alternative way of looking at streaky hitting performances uses batting averages computed over short time intervals. One may be interested in exploring DiMaggio's batting average in this manner. He must have been a hot hitter during his 56-game hitting streak, and perhaps DiMaggio was somewhat cold in other periods during the season.

In general, suppose we are interested in computing a player's batting average over a width (or window) of 10 games. We want to compute the batting average over games 1 to 10, over games 2 to 11, over games 3 to 12, and so on. These batting averages would be the sum of hits divided by the sum of at-bats over the 10-game periods. These short-term batting averages are commonly called moving averages.

The function `moving_average()` below computes these moving averages. The arguments to the function are a data frame with variables `H` and `AB`, and the window of games `width`. The main tools in this function are the functions `rollmean()` and `rollsum()` from the `zoo` package. The output of the function is a data frame with two variables (justifying the use of `transmute()` in place of `mutate()`): `Game` and `Average`. The variable `Game` gives the game number value in the middle of the window, and `Average` is the corresponding batting average over the game window.

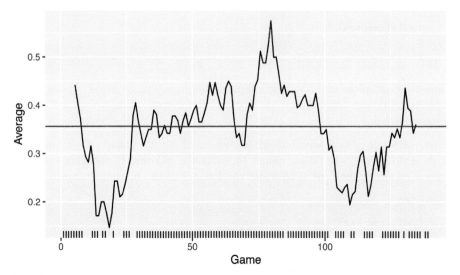

FIGURE 10.1
Moving average plot of DiMaggio's batting average for the 1941 season using a window of 10 games. The horizontal line shows DiMaggio's season batting average. The games where DiMaggio had at least one base hit are displayed on the horizontal axis.

```
library(zoo)
moving_average <- function(df, width) {
  N <- nrow(df)
  df |>
    transmute(
      Game = rollmean(1:N, k = width, fill = NA),
      Average = rollsum(H, width, fill = NA) /
        rollsum(AB, width, fill = NA)
    )
}
```

After the function `moving_average()` is read into R, it is easy to compute DiMaggio's batting average over short time intervals. Suppose we consider a window of 10 games. In the following code, we use `moving_average()` to compute the moving batting averages and pass the output to `ggplot()` and `geom_line()` to construct a line graph of these averages (see Figure 10.1). We add a horizontal line using the `geom_hline()` function at DiMaggio's season batting average so one can easily see when Joe was relatively hot and cold during the season. To relate this display with DiMaggio's hitting streaks, we use the `geom_rug()` function to display the games where Joe had at least one hit on the horizontal axis.

```
joe_ma <- moving_average(joe, 10)

ggplot(joe_ma, aes(Game, Average)) +
  geom_line() +
  geom_hline(
    data = summarize(joe, bavg = sum(H)/sum(AB)),
    aes(yintercept = bavg), color = "red"
  ) +
  geom_rug(
    data = filter(joe, had_hit == 1),
    aes(Rk, .3 * had_hit), sides = "b",
    color = crcblue
  )
```

This figure dramatically shows that DiMaggio's hitting performance climbed steadily during his 56-game hitting streak and he actually had a short-term 10-game batting average over .500 during the streak. DiMaggio had a noticeable hitting slump in the second half of the season and he hit bottom about Game 110. In practice, the appearance of this graph may depend on the choice of time interval (argument `width` in the function `moving_average()`) and one should experiment with several `width` choices to get a better understanding of a hitter's short-term batting performance.

10.3 Streaks in Individual At-Bats

The previous section considered hitting streaks at a game-to-game level. Since records of individual plate appearances are available in the Retrosheet play-by-play files, it is straightforward to explore hitting streaks at this finer level. Ichiro Suzuki was one of the most exciting hitters in baseball, especially for his ability to hit singles, many of the infield variety. We explore the streakiness patterns in Suzuki's play-by-play hitting data for the 2016 season.

We begin by reading the Retrosheet play-by-play file for the 2016 season, storing the file in the data frame `retro2016`.

```
retro2016 <- read_rds(here::here("data/retro2016.rds"))
```

We use the `filter()` function to define a new data frame `ichiro_AB`; records are chosen where the batting id is `suzui001` (Suzuki's code id) and the at-bat flag is `TRUE`. (In this exploration, only Suzuki's official at-bats are considered.)

```
ichiro_AB <- retro2016 |>
  filter(bat_id == "suzui001", ab_fl == TRUE)
```

10.3.1   Streaks of hits and outs

We record each at-bat if a hit occurred. There is a variable `h_fl` in the Retrosheet data recording the number of bases for a hit. Using the `if_else()` function, we define a new variable H that is 1 if a hit occurs and 0 otherwise. To make sure that these at-bats are correctly ordered in time during the season, we define a variable `date` (extracted from the `game_id` variable using the `str_sub()` function), and the `arrange()` function sorts the data frame `ichiro_AB` by date.

```
ichiro_AB <- ichiro_AB |>
  mutate(
    H = if_else(h_fl > 0, 1, 0),
    date = str_sub(game_id, 4, 12),
    AB = 1
  ) |>
  arrange(date)
```

From the variable H, we identify the lengths of all hitting streaks, where a streak refers to a sequence of consecutive base hits. Using the `streaks()` function defined in Section 10.2 and filtering by `values == 1`, we obtain the streak lengths for Suzuki in the 2016 season.

```
ichiro_AB |>
  pull(H) |>
  streaks() |>
  filter(values == 1) |>
  pull(lengths)
```

```
 [1] 1 1 2 1 2 1 1 1 1 1 1 1 2 5 1 3 1 1 1 1 2 2 1 3 1 1 3 1 1 2 2
[31] 1 1 2 1 1 1 1 1 1 2 1 1 1 1 1 1 1 2 1 1 2 1 1 1 1 1 1 1 1 3
[61] 1 1 1 1 1 2 2 1 1 1
```

As expected, most of the hitting streaks lengths are 1, although once Suzuki had five consecutive hits.

It may be more interesting to explore the lengths of the gaps between hits. We apply the function `streak()` a second time filtering by `values == 0` to find the lengths of all of the gaps between hits that are 1 or larger.

```
ichiro_out <- ichiro_AB |>
  pull(H) |>
  streaks() |>
  filter(values == 0)
ichiro_out |>
  pull(lengths)
```

```
[1]   2  1  2  1  4  2  5  2  2  3  4  3  1  1  1  1 11 12  1  2
[21]  7  3  1  1  2  2  1  1  3  4  1  4  8  1  2  1  4  2  1  1
[41]  4  2  7  1 11  4  3  1 10  1  3  1 11  8  1  3  6  5  1  3
[61]  3  1  1  3  2  1  1 18  2  3
```

This output is more interesting. We construct a frequency table of this output by use of the group_by() and count() functions.

```
ichiro_out |>
  group_by(lengths) |>
  count()
```

```
# A tibble: 12 x 2
# Groups:   lengths [12]
   lengths      n
     <int> <int>
1        1    26
2        2    13
3        3    11
4        4     7
5        5     2
6        6     1
7        7     2
8        8     2
9       10     1
10      11     3
11      12     1
12      18     1
```

We see that Suzuki had a streak of 18 outs once, a streak of 12 outs twice, and a streak of 11 outs three times.

10.3.2 Moving batting averages

Another way to view Suzuki's streaky batting performance is to consider his batting average over short time intervals, analogous to what we did for DiMaggio for his game-to-game hitting data. Using the moving_average() function, we construct a moving average plot of Ichiro's batting average using

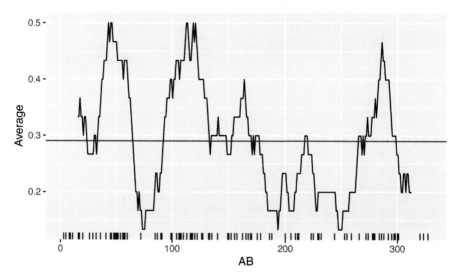

FIGURE 10.2
Moving average plot of Ichiro Suzuki's batting average for the 2016 season
using a window of 30 at-bats. The horizontal line shows Suzuki's season batting
average. The at-bats where Suzuki had at least one base hit are shown on the
horizontal axis.

a window of 30 at-bats (see Figure 10.2). Using the `geom_rug()` function, we
display the at-bats where Ichiro had hits. The long streaks of outs are visible
as gaps in the rug plot. During the middle of the season, Ichiro had a 30 at-bat
batting average exceeding 0.500, while during other periods, his 30 at-bat
average was as low as 0.100.

```
ichiro_H <- ichiro_AB |>
  mutate(AB_Num = row_number()) |>
  filter(H == 1)

moving_average(ichiro_AB, 30) |>
  ggplot(aes(Game, Average)) +
  geom_line() + xlab("AB") +
  geom_hline(yintercept = mean(ichiro_AB$H),
             color = "red") +
  geom_rug(
    data = ichiro_H,
    aes(AB_Num, .3 * H), sides = "b",
    color = crcblue
  )
```

10.3.3 Finding hitting slumps for all players

In our exploration of Suzuki's batting performance, we saw that he had a "0 for 18" hitting performance during the season. Should we be surprised by a hitting slump of length 18? Let's compare Suzuki's long slump with the longest slumps for all regular players during the 2016 season.

First, we write a new function `longest_ofer()` that computes the length of the longest hitting slump for a given batter. (An "ofer" is a slang word for a hitless streak in baseball.) The input to this function is the batter id code `batter` and the output of the function is the length of the longest slump.

```
longest_ofer <- function(batter) {
  retro2016 |>
    filter(bat_id == batter, ab_fl == TRUE) |>
    mutate(
      H = ifelse(h_fl > 0, 1, 0),
      date = substr(game_id, 4, 12)
    ) |>
    arrange(date) |>
    pull(H) |>
    streaks() |>
    filter(values == 0) |>
    summarize(max_streak = max(lengths))
}
```

After reading this function into R, we confirm that it works by finding the longest hitting slump for Suzuki.

```
longest_ofer("suzui001")
```

```
# A tibble: 1 x 1
  max_streak
       <int>
1         18
```

Suppose we want to compute the length of the longest hitting slump for all players in this season with at least 400 at-bats. Using the `group_by()` and `summarize()` functions, we compute the number of at-bats for all players, and `players_400` contains the id codes of all players with 400 or more at-bats. By use of the `map()` function together with the new `longest_ofer()` function, we compute the length of the longest slump for all regular hitters. The final object `reg_streaks` is a data frame with variables `bat_id` and `max_streak`.

```
players_400 <- retro2016 |>
  group_by(bat_id) |>
```

```
    summarize(AB = sum(ab_fl)) |>
    filter(AB >= 400) |>
    pull(bat_id)
  reg_streaks <- players_400 |>
    set_names() |>
    map(longest_ofer) |>
    list_rbind() |>
    mutate(bat_id = players_400)
```

To decipher the player ids, it is helpful to merge the data frame of the longest hitting slumps `reg_streaks` with the player roster information contained in the `People` data frame from the **Lahman** package. We then apply the `inner_join()` function, merging data frames `reg_streaks` and the `People` data frame, matching on the variables `bat_id` (in `reg_streaks`) and `retroID` (in `People`). The rows of the resulting data frame are reordered using the slump lengths in decreasing order using the function `arrange()` with the `desc()` modifier. The top six slump lengths are displayed by the `slice_head()` function below.

```
  library(Lahman)
  reg_streaks |>
    inner_join(People, by = c("bat_id" = "retroID")) |>
    mutate(Name = paste(nameFirst, nameLast)) |>
    arrange(desc(max_streak)) |>
    select(Name, max_streak) |>
    slice_head(n = 6)
```

```
# A tibble: 6 x 2
  Name              max_streak
  <chr>                  <int>
1 Carlos Beltran            32
2 Denard Span               30
3 Brandon Moss              29
4 Eugenio Suarez            28
5 Francisco Lindor          27
6 Albert Pujols             26
```

The six longest hitting slumps during the 2016 season were by Carlos Beltran (32), Denard Span (30), Brandon Moss (29), Eugenio Suarez (28), Francisco Lindor (27) and Albert Pujols (26). Relative to these long hitting slumps, Suzuki's hitting slump of 18 at-bats looks short.

10.3.4 Were Ichiro Suzuki and Mike Trout unusually streaky?

In the previous section, patterns of streakiness of hit/out data were compared for all players in the 2016 season. An alternative way to look at the streakiness of a player is to contrast his streaky pattern of hitting with streaky patterns under a "random" model.

To illustrate this method, consider a hypothetical player who bats 13 times with the outcomes

$$0, 1, 0, 0, 1, 1, 0, 0, 0, 0, 1, 1, 1.$$

We define a measure of streakiness based on this sequence of hits and outs. One good measure of streakiness or clumpiness in the sequence is the sum of squares of the gaps between successive hits. In this example, the gaps between hits are 1, 2, and 4, and the sum of squares of the gaps is $S = 1^2 + 2^2 + 4^2 = 21$.

Is the value of streakiness statistic $S = 21$ large enough to conclude that this player's pattern of hitting is non-random? We answer this question by a simple simulation experiment. If the player sequence of hit/out outcomes is truly random, then all possible arrangements of the sequence of 6 hits and 7 outs are equally likely. We randomly arrange the sequence 0, 1, 0, 0, 1, 1, 0, 0, 0, 0, 1, 1, 1, find the gaps, and compute the streakiness measure S. This randomization procedure is repeated many times, collecting, say, 1000 values of the streakiness measure S. We then construct a histogram of the values of S—this histogram represents the distribution of S under a random model. If the observed value of $S = 21$ is in the middle of the histogram, then the player's pattern of streakiness is consistent with a random model. On the other hand, if the value $S = 21$ is in the right tail of this histogram, then the observed streaky pattern is not consistent with "random" streakiness and there is evidence that the player's pattern of hits and outs is non-random.

We first illustrate this method for Ichiro Suzuki's 2016 hitting data.

The clumpiness or streakiness is measured by the sum of squares of all gaps between hits. We use the function `streaks()` to find all of the gaps and by filtering for `values == 0`, and we focus on the gaps between successive hits. Each of the gap values is squared and the `sum()` function computes the sum.

```
ichiro_S <- ichiro_AB |>
  pull(H) |>
  streaks() |>
  filter(values == 0) |>
  summarize(C = sum(lengths ^ 2)) |>
  pull()
ichiro_S
```

`[1] 1532`

The value of Suzuki's streakiness statistic S is 1532.

Next, we write a function `random_mix()` to perform one iteration of the simulation experiment where the input y is a vector of 0s and 1s. The `sample()` function finds a random arrangement of y, the `streaks()` function with the

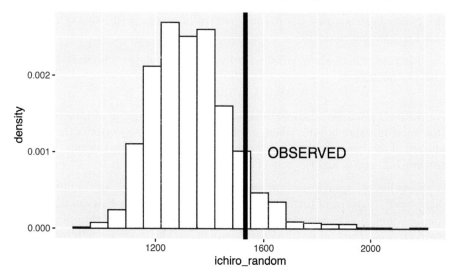

FIGURE 10.3
Histogram of one thousand values of the clumpiness statistic assuming all arrangements of hits and outs for Suzuki are equally likely. The observed value of the clumpiness statistic for Suzuki is shown using a vertical line.

restriction `values == 0` finds the gaps between hits, and the sum of squares of the gaps is computed.

```
random_mix <- function(y) {
  y |>
    sample() |>
    streaks() |>
    filter(values == 0) |>
    summarize(C = sum(lengths ^ 2)) |>
    pull()
}
```

We repeat this simulation experiment 1000 times using the `map_int()` function, and store the values of the streakiness statistic in the vector `ichiro_random`.

```
ichiro_random <- 1:1000 |>
  map_int(~random_mix(ichiro_AB$H))
```

We construct a histogram of the values of `ichiro_random` using the `geom_hist()` function, and use the `geom_vline()` function to overlay the clumpiness value (1532) for Suzuki (see Figure 10.3).

```
ggplot(enframe(ichiro_random), aes(ichiro_random)) +
  geom_histogram(
    aes(y = after_stat(density)), bins = 20,
    color = crcblue, fill = "white"
  ) +
  geom_vline(xintercept = ichiro_S, linewidth = 2) +
  annotate(
    geom = "text", x = ichiro_S * 1.15,
    y = 0.0010, label = "OBSERVED", size = 5
  )
```

Since the value of 1532 is in the right tail of the histogram distribution, the streakiness pattern in Suzuki's hitting is not so consistent with a random model (above the 90% percentile, as computed with the `quantile()` function). There is some evidence that Suzuki was truly streaky in his hitting during the 2016 season.

```
quantile(ichiro_random, probs = 0:10/10)
```

```
  0%   10%   20%   30%   40%   50%   60%   70%   80%   90%  100%
 948  1160  1208  1246  1280  1319  1358  1396  1448  1518  2202
```

This method can be used to check if the streaky patterns of any hitter are non-random. We construct a new function `clump_test()` using the R code previous discussed. The input is the player id code `playerid` and the season batting data frame `data`. One thousand values of the clumpiness measure are computed by 1000 replications of the simulation procedure. A histogram of the clumpiness measures is constructed and the observed clumpiness statistic is shown as a vertical line.

```
clump_test <- function(data, playerid) {
  player_ab <- data |>
    filter(bat_id == playerid, ab_fl == TRUE) |>
    mutate(
      H = ifelse(h_fl > 0, 1, 0),
      date = substr(game_id, 4, 12)
    ) |>
    arrange(date)

  stat <- player_ab |>
    pull(H) |>
    streaks() |>
    filter(values == 0) |>
    summarize(C = sum(lengths ^ 2)) |>
    pull()
```

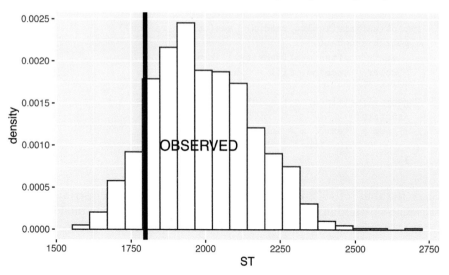

FIGURE 10.4
Histogram of one thousand values of the clumpiness statistic assuming all arrangements of hits and outs for the 2016 Mike Trout are equally likely. The observed value of the clumpiness statistic for Trout is shown using a vertical line.

```
ST <- 1:1000 |>
  map_int(~random_mix(player_ab$H))

ggplot(enframe(ST), aes(ST)) +
  geom_histogram(
    aes(y = after_stat(density)), bins = 20,
    color = crcblue, fill = "white"
  ) +
  geom_vline(xintercept = stat, linewidth = 2) +
  annotate(
    geom = "text", x = stat * 1.10,
    y = 0.0010, label = "OBSERVED", size = 5
  )
}
```

Was Mike Trout streaky during the 2016 season? To investigate the non-randomness of Trout's sequence of hit/out data, we run the function `clump_test()` using Trout's player id code **troum001** and show the resulting histogram display in Figure 10.4.

```
clump_test(retro2016, "troum001")
```

Note that Trout's clumpiness measure is in the left tail of this distribution, indicating that Trout did not display more streakiness than one would expect by chance.

10.4 Local Patterns of Statcast Launch Velocity

In our discussion of hitting slumps and streaks, our focus is on either getting a hit or an out in an official at-bat. With the new Statcast data, one can propose alternative definitions of a successful plate appearance from measurements from the ball that is put in-play. In particular, we explore patterns of slumps and streaks for a given player using the launch velocity of batted balls.

In the following code, we read in the file `sc_2017_ls.rds` that contains data on every pitch in the 2017 season. To focus on batted balls, the `filter()` function is used to select the pitches where the variable `type` is equal to X. By use of the `group_by()` and `summarize()` functions, we return a data frame `launch_speeds` that contains the number of batted balls and the sum of the launch speeds for each player for each game.

```
sc_2017_ls <- read_rds(here::here("data/sc_2017_ls.rds"))
sc_ip2017 <- sc_2017_ls |>
  filter(type == "X")
launch_speeds <- sc_ip2017 |>
  group_by(player_name, game_date) |>
  arrange(game_date) |>
  summarize(
    bip = n(),
    sum_LS = sum(launch_speed)
  )
```

Here we focus on players who had at least 250 batted balls. Below we compute the number of batted balls for each player, merge this information with the data frame `launch_speeds`, and use the `filter()` function to create a new data frame `ls_250` containing the game-to-game launch speed data for the regular players.

```
ls_250 <- sc_ip2017 |>
  group_by(player_name) |>
  summarize(total_bip = n()) |>
  filter(total_bip >= 250) |>
  inner_join(launch_speeds, by = "player_name")
```

Say we are interested in looking at a player's mean launch speed over groups of five games—games 1-5, games 6-10, games 11-15, and so on. The collection of player's launch speeds for all games in a season is represented as a data frame, where rows correspond to games, the sum_LS column corresponds to the sum of launch speeds for these games, and the BIP column corresponds to the number of batted balls. The new function regroup() collapses a player's batting performance matrix into groups of size group_size, where a particular row will correspond to the sum of launch speeds and sum of the count of batted balls in a particular group of games. (In our exploration, we use groups of size 5.)

```
regroup <- function(data, group_size) {
  out <- data |>
    mutate(
      id = row_number() - 1,
      group_id = floor(id / group_size)
    )
  # hack to avoid a small leftover bin!
  if (nrow(data) %% group_size != 0) {
    max_group_id <- max(out$group_id)
    out <- out |>
      mutate(
        group_id = if_else(
          group_id == max_group_id, group_id - 1, group_id
        )
      )
  }
  out |>
    group_by(group_id) |>
    summarize(
      G = n(), bip = sum(bip), sum_LS = sum(sum_LS)
    )
}
```

To illustrate this grouping operation, we collect the game-to-game hitting data for A.J. Pollock in the data frame aj. As before, to make sure the data is chronologically ordered, the rows are ordered by increasing values of game_date using arrange(). We then apply the regroup() function to the data frame aj. The output is a data frame with four columns: the first column contains the group id, the second contains the number of games in each group, the third is the number of batted balls for each group of five games, and the last column contains the sum of launch velocities. (Only the first few rows of this data frame are displayed.)

```
aj <- ls_250 |>
  filter(player_name == "A.J. Pollock") |>
  arrange(game_date)
aj |>
```

```
regroup(5)   |>
slice_head(n = 5)
```

```
# A tibble: 5 x 4
  group_id      G   bip  sum_LS
     <dbl> <int> <int>   <dbl>
1        0     5    21   1850.
2        1     5    17   1500.
3        2     5    15   1376.
4        3     5    18   1598.
5        4     5    17   1505.
```

We have illustrated the process of finding the five-game hitting data for A.J. Pollock. When we look at the sequence of five-game launch speed data for an arbitrary player, the mean launch speeds for a consistent player will have small variation, and the values for a streaky player will have high variability. A common measure of variability is the standard deviation, the average size of the deviations from the mean.

We write a new function to compute the mean and standard deviation of the grouped launch speed means for a given player. This function `summarize_streak_data()` performs this operation for the game-by-game data frame of launch speeds `ls_250`, a given player with name `name`, and a grouping of `group_size` games (by default 5}). The output is a vector with the number of batted balls, the mean of the group mean launch speeds `Mean` and the standard deviation of the mean launch speeds `SD`.

```
summarize_streak_data <- function(data, name, group_size = 5) {
  data |>
    filter(player_name == name) |>
    arrange(game_date) |>
    regroup(group_size) |>
    summarize(
      balls_in_play = sum(bip),
      Mean = mean(sum_LS / bip, na.rm = TRUE),
      SD = sd(sum_LS / bip, na.rm = TRUE)
    )
}
```

To illustrate the use of this function, we apply it to A. J. Pollock's hitting data.

```
aj_sum <- summarize_streak_data(ls_250, "A.J. Pollock")
aj_sum
```

```
# A tibble: 1 x 3
```

```
balls_in_play  Mean    SD
        <int> <dbl> <dbl>
1          354  87.9  3.44
```

Pollock had 354 batted balls, the mean of his five-game launch speed means was 87.9 and the standard deviation of his five-game launch speed means was 3.44.

We now apply the function `summarize_streak_data()` to all players with at least 250 batted balls in the 2017 season. We define the vector `player_list` to be the vector of all unique player ids and use the `map()` function to apply `summarize_streak_data()` to all players in `player_list`.

```
player_list <- ls_250 |>
  pull(player_name) |>
  unique()
results <- player_list |>
  map(summarize_streak_data, data = ls_250) |>
  list_rbind() |>
  mutate(Player = player_list)
```

We construct a scatterplot of the means and standard deviations of the mean launch speeds of these "regular" players in Figure 10.5. By use of the `geom_label_repel()` function, we label with player names the points corresponding to the largest and smallest standard deviations.

```
library(ggrepel)
ggplot(results, aes(Mean, SD)) +
  geom_point() +
  geom_label_repel(
    data = filter(results, SD > 5.63 | SD < 2.3 ),
    aes(label = Player)
  )
```

The streakiest hitter during the 2017 season using this standard deviation measure was Michael Conforto. Conversely, the most consistent player, Dexter Fowler, is identified as the one with the smallest standard deviation of the five-game mean launch speeds. These two players can be compared graphically by plotting their five-game launch speed values against the period number (see Figure 10.6).

We create a new function `get_streak_data()` to compute the vector of five-game launch speed means for a particular player. This function is a simple modification of the function `summarize_streak_data()` where the period number `Period` and mean launch speed `launch_speed_avg` are computed for each five-game period.

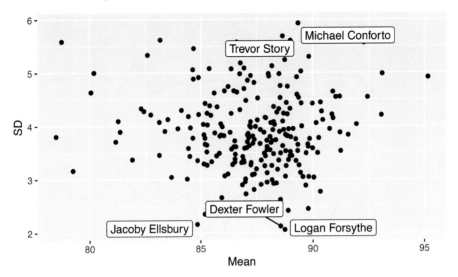

FIGURE 10.5
Scatterplot of means and standard deviations of the five-game averages of launch speeds of regular players during the 2017 season. The labeled points correspond to the players with the smallest and largest standard deviations, corresponding to consistent and streaky hitters.

```
get_streak_data <- function(data, name, group_size = 5) {
  data |>
    filter(player_name == name) |>
    arrange(game_date) |>
    regroup(group_size) |>
    mutate(
      launch_speed_avg = sum_LS / bip,
      Period = row_number()
    )
}
```

Using this new function, we create a data frame `streaky` with Conforto and Fowler's streakiness data. First we use the `set_names()` function to build a named vector of players. Then we use the `map()` function to apply `get_streak_data()` to each player in the vector.

The graphics functions `ggplot()`, `geom_line()`, and `facet_wrap()` in the **ggplot2** package are used to create the line graphs. One nice feature of **ggplot2** graphics is that it automatically uses the same vertical scale for the two panels and shows the player names on the right of the graph.

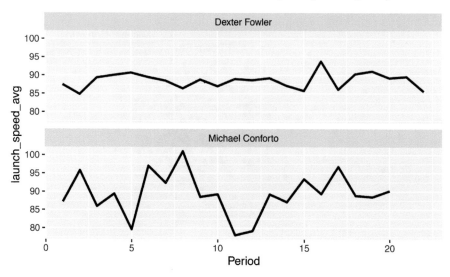

FIGURE 10.6
Line plots of five-game average launch velocities of Michael Conforto and
Dexter Fowler for the 2017 season. Conforto had a streaky pattern of launch
velocities and Fowler's pattern is very consistent.

```
streaky <- c("Michael Conforto", "Dexter Fowler") |>
  set_names() |>
  map(get_streak_data, data = ls_250) |>
  list_rbind(names_to = "Player")

ggplot(streaky, aes(Period, launch_speed_avg)) +
  geom_line(linewidth = 1) +
  facet_wrap(vars(Player), ncol = 1)
```

Note that, as expected, Conforto and Fowler have dramatically different patterns of five-game launch speed means. Most of Fowler's five-game mean launch
speeds fall between 85 and 90 mph. In contrast, Conforto had a change in
mean launch speed from 80 to 100 mph in two periods; he was a remarkably
streaky hitter during the 2017 season.

10.5   Further Reading

There is much interest in streaky performances of baseball players in the
literature. Gould (1989), Berry (1991), and Seidel (2002) discuss the significance

of DiMaggio's hitting streak in the 1941 season. Albert and Bennett (2003), Chapter 5, describes the difference between observed streakiness and true streakiness and give an overview of different ways of detecting streakiness of hitters. Albert (2008) and McCotter (2010) discuss the use of randomization methods to detect if there is more streakiness in hitting data than one would expect by chance.

10.6 Exercises

1. Ted Williams

The data file `williams.1941.csv` contains Ted Williams game-to-game hitting data for the 1941 season. This season was notable in that Williams had a season batting average of .406 (the most recent season batting average exceeding .400). Read this dataset into R.

a. Using the R function `streaks()`, find the lengths of all of Williams' hitting streaks during this season. Compare the lengths of his hitting streaks with those of Joe DiMaggio during this same season.

b. Use the function `streaks()` to find the lengths of all hitless streaks of Williams during the 1941 season. Compare these lengths with those of DiMaggio during the 1941 season.

2. Ted Williams, Continued

a. Use the R function `moving_average()` to find the moving batting averages of Williams for the 1941 season using a window of 5 games. Graph these moving averages and describe any hot and cold patterns in Williams hitting during this season.

b. Compute and graph moving batting averages of Williams using several alternative choices for the window of games.

3. Streakiness of the 2008 Lance Berkman

Lance Berkman had a remarkable hot period of hitting during the 2008 season.

a. Download the Retrosheet play-by-play data for the 2008 season, and extract the hitting data for Berkman.

b. Using the function `streaks()`, find the lengths of all hitting streaks of Berkman. What was the length of his longest streak of consecutive hits?

c. Use the `streaks()` function to find the lengths of all streaks of consecutive outs. What was Berkman's longest "ofer" during this season?

d. Construct a moving batting average plot using a window of 20 at-bats. Comment on the patterns in this graph; was there a period when Berkman was unusually hot?

4. Streakiness of the 2008 Lance Berkman, Continued

a. Use the method described in Section 10.3.4 to see if Berkman's streaky patterns of hits and outs are consistent with patterns from a random model.

b. The method of Section 10.3.4 used the sum of squares of the gaps as a measure of streakiness. Suppose one uses the longest streak of consecutive outs as an alternative measure. Rerun the method with this new measure and see if Berkman's longest streak of outs is consistent with the random model.

5. Streakiness of All Players During the 2008 Season

a. Using the 2008 Retrosheet play-by-play data, extract the hitting data for all players with at least 400 at-bats.

b. For each player, find the length of the longest streak of consecutive outs. Find the hitters with the longest streaks and the hitters with shortest streaks. How does Berkman's longest "oh-for" compare in the group of longest streaks?

6. Streakiness of All Players During the 2008 Season, Continued

a. For each player and each game during the 2008 season, compute the sum of $wOBA$ weights and the number of plate appearances PA (see Section 10.4).

b. For each player with at least 500 PA, compute the $wOBA$ over groups of five games (games 1-5, games 6-10, etc.) For each player, find the standard deviation of these five-game $wOBA$, and find the ten most streaky players using this measure.

7. The Great Streak

The Retrosheet website recently added play-by-play data for the 1941 season when Joe DiMaggio achieved his 56-game hitting streak.

a. Download the 1941 play-by-play data from the Retrosheet website.

b. Confirm that DiMaggio had three "0 for 12" streaks during the 1941 season.

c. Use the method described in Section 10.3.4 to see if DiMaggio's streaky patterns of hits and outs in individual at-bats are consistent with patterns from a random model.

d. DiMaggio is perceived to be very streaky due to his game-to-game hitting accomplishment during the 1941 season. Based on your work, is DiMaggio's pattern of hitting also very streaky on individual at-bats?

11

Using a Database to Compute Park Factors

11.1 Introduction

Thus far in this book, we have performed analyses entirely from baseball datasets loaded into R. That was possible because we were dealing with datasets with a relatively small number of rows. However, when one wants to work on multiple seasons of play-by-play (or pitch-by-pitch) data, it becomes more difficult to manage all of the data inside R.[1] While Retrosheet game logs consist of approximately 250,000 records, there are more than 10 million Retrosheet play-by-play events, and Statcast provides data on roughly 800,000 pitches per year for MLB games.

A solution to this *big data* problem is to store them in a *Relational Database Management System* (RDBMS), connect it to R, and access only the data needed for the particular analysis. In this chapter, we provide some guidance on this approach. Our choice for the RDBMS is *MariaDB*, a fork of *MySQL*, which is likely the most popular open-source RDBMS[2]. However, readers familiar with other software (e.g., *PostgreSQL*, *SQLite*) can find similar solutions for their RDBMS of choice. We compare alternative big data strategies in Chapter 12.

Here, we use MySQL to gain some understanding of park effects in baseball. Unlike most other team sports, in baseball the size and configuration of the playing surface varies greatly across ballparks. The left-field wall in Fenway Park (home of the Boston Red Sox) is listed at 310 feet from home plate, while the left-field fence in Wrigley Field (home of the Chicago Cubs) is 355 feet away. The left-field wall in Boston, commonly known as The Green Monster, is 37 feet high, while the left-field fence in Dodger Stadium in Los Angeles is only four feet high. Such differences in ballpark shapes and dimensions and

---

[1] R by default reads data into memory (RAM), thus imposing limits on the size of datasets it can read.

[2] MariaDB is designed to be a drop-in replacement for MySQL. In fact, the MariaDB application is called `mysql`. In some cases, we may use the terms interchangeably.

DOI: 10.1201/9781032668239-11   257

the prevailing weather conditions have a profound effect on the game and the associated measures of player performance.

We first show how to obtain and set up a MariaDB Server, and then illustrate connecting R to the database for the purpose of first inserting, and then retrieving data. This interface is used to present evidence of the effect of Coors Field (home of the Colorado Rockies) on run scoring. We direct the reader to online resources providing baseball data (of the seasonal and pitch-by-pitch types) ready for import into MySQL. We conclude the chapter by providing readers with a basic approach for calculating park factors and using these factors to make suitable adjustments to players' stats.

11.2   Installing MySQL and Creating a Database

Since this is a book focusing on R, we emphasize the use of MariaDB together with R. A user can install a MariaDB Server on her own from https://mari adb.com/downloads/. In this third edition of the book, we will demonstrate the use of *MariaDB*, which is an open-source fork of MySQL that is almost totally compatible. Somewhat confusingly, MariaDB installs a server that is called MySQL. Readers should be able to follow these instructions[3] whether they are using MariaDB or MySQL.

Many people already have a MySQL server running on their computers. The easiest way to see whether this is the case is to check for a running process using a terminal.

```
ps -ax | grep "mariadb"
```

```
404058 ?        SNsl   2:04 /usr/sbin/mariadbd
490520 pts/0    SN+    0:00 sh -c 'ps' -ax | grep 'mariadb'
490522 pts/0    SN+    0:00 grep mariadb
```

Here, the process at /usr/sbin/mariadbd is the MySQL server.

Once the MySQL server is running, you will connect to it and create a new database. We illustrate how this can be achieved using the command line MySQL client. In this case, we log into MySQL as the user root and supply the corresponding password. To replicate the work in this chapter, you will need access to an account on the server that has sufficient privileges to create new users and databases. Please consults the MariaDB documentation if you

---

[3]Appendix F of Benjamin S. Baumer, Kaplan, and Horton (2021b) (available at https://mdsr-book.github.io/mdsr3e/F-dbsetup.html) also provides step-by-step instructions for setting up a SQL server.

run into trouble.

```
mysql -u root -p
```

Once inside MySQL, we can create a new database called `abdwr` with the following command.

```
CREATE DATABASE abdwr;
```

Similarly, we create a new user `abdwr` that uses the password `spahn`.

```
CREATE USER 'abdwr'@'localhost' IDENTIFIED BY 'spahn';
```

Next, we give the user `abdwr` all the privileges on the database `abdwr`, and then force the server to reload the privileges database.

```
GRANT ALL ON abdwr.* TO 'abdwr'@'localhost' WITH GRANT OPTION;
FLUSH PRIVILEGES;
```

11.2.1 Setting up an Option File

In the previous section, we created a user named `abdwr` on our MariaDB server and gave that user the password `spahn`. In general, exposing passwords in plain text is not a good idea, and we will now demonstrate how to use an option file to store MariaDB/MySQL database credentials.

An option file is just a plain text file with database connection options embedded in it. If this file is stored at `~/.my.cnf` it will be automatically read when you try to log in to any MariaDB/MySQL database, without the need for you to type in your password (or any other connection parameters).

The option file you need for this book looks like this:

```
cat ~/.my.cnf
```

```
[abdwr]
database="abdwr"
user="abdwr"
password="spahn"
```

With this option file, you can connect to the MariaDB server via the command line with just:

```
mysql
```

11.3   Connecting R to MySQL

Connections to SQL databases from R are best managed through the **DBI** package, which provides a common interface to many different RDBMSs. Many **DBI**-compliant packages provide direct connections to various databases. For example, the **RMariaDB** package connects to MariaDB databases, the **RPostgreSQL** package connects to PostgreSQL databases, the **RSQLite** package connects to SQLite databases, and the **odbc** package connects to any database that supports *ODBC*. Because of subtle changes in licensing, since the second edition of this book development of database tools to connect to MySQL servers has moved from the **RMySQL** to the **RMariaDB** package. Please see https://solutions.posit.co/connections/db/ for the most current information about connecting your database of choice to R.

The **RMariaDB** package provides R users with functions to connect to a MySQL database. Readers who plan to make extensive use of MySQL connections are encouraged to make the necessary efforts for installing **RMariaDB**.

11.3.1   Connecting using `RMariaDB`

The **DBI** function `dbConnect()` creates a connection to a database server, and returns an object that stores the connection information.

If you don't care about password security, you can explicitly state those parameters as in the code below. The `user` and `password` arguments indicate the user name and the password for accessing the MySQL database (if they have been specified when the database was created), while `dbname` indicates the default database to which R will be connected (in our case `abdwr`, created in Section 11.2).

```
library(RMariaDB)
con <- dbConnect(
  MariaDB(), dbname = "abdwr",
  user = "abdwr", password = "spahn"
)
```

Alternatively, we could connect using the `group` argument along with the option file described Section 11.2.1.

```
library(RMariaDB)
con <- dbConnect(MariaDB(), group = "abdwr")
```

Note that the connection is assigned to an R object (`con`), as it will be required as an argument by several functions.

```
class(con)
```

```
[1] "MariaDBConnection"
attr(,"package")
[1] "RMariaDB"
```

To remove a connection, use `dbDisconnect()`.

11.3.2   Connecting R to other SQL backends

The process of connecting R to any other SQL backend is very similar to the one outlined above for MySQL. Since the connections are all managed by **DBI**, one only needs to change the database backend. For example, to connect to a PostgreSQL server instead of MySQL, one loads the **RPostgreSQL** package instead of **RMariaDB**, and uses the `PostgreSQL()` function instead of the `MariaDB()` function in the call to `dbConnect()`. The rest of the process is the same. The resulting PostgreSQL connection can be used in the same way as the MySQL connection is used below.

11.4   Filling a MySQL Game Log Database from R

The game log data files are currently available at the Retrosheet web page https://www.retrosheet.org/gamelogs/index.html. By clicking on a single year, one obtains a compressed (`.zip`) file containing a single text file of that season's game logs. Here we first create a function for loading a season of game logs into R, then show how to append the data into a MySQL table. We then iterate this process over several seasons, downloading the game logs from Retrosheet and appending them to the MySQL table.

Since game log files downloaded from Retrosheet do not have column headers, the resulting `gl2012` data frame has meaningful names stored in the `game_log_header.csv` file as done elsewhere in this book.

11.4.1   From Retrosheet to R

The `retrosheet_gamelog()` function, shown below, takes the year of the season as input and performs the following operations:

- Imports the column header file
- Downloads the zip file of the season from Retrosheet
- Extracts the text file contained in the downloaded zip file
- Reads the text file into R, using the known column headers
- Removes both the compressed and the extracted files

- Returns the resulting data frame

```
retrosheet_gamelog <- function(season) {
  require(abdwr3edata)
  require(fs)
  dir <- tempdir()
  glheaders <- retro_gl_header
  remote <- paste0(
    "http://www.retrosheet.org/gamelogs/gl",
    season,
    ".zip"
  )
  local <- path(dir, paste0("gl", season, ".zip"))
  download.file(url = remote, destfile = local)
  unzip(local, exdir = dir)
  local_txt <- gsub(".zip", ".txt", local)
  gamelog <- here::here(local_txt) |>
    read_csv(col_names = names(glheaders))
  file.remove(local)
  file.remove(local_txt)
  return(gamelog)
}
```

After the function has been read into R, one season of game logs (for example, the year 2012) can be read into R by typing the command:

```
gl2012 <- retrosheet_gamelog(2012)
```

11.4.2   From R to MySQL

Next, we transfer the data in the gl2012 data frame to the abdwr MySQL database. In the lines that follow, we use the dbWriteTable() function to append the data to a table in the MySQL database (which may or may not exist).

Here are some notes on the arguments in the dbWriteTable() function:

- The conn argument requires an open connection; here the one that was previously defined (con) is specified.
- The name argument requires a string indicating the name of the table (in the database) where the data are to be appended.
- The value argument requires the name of the R data frame to be appended to the table in the MySQL database.
- Setting append to TRUE indicates that, should a table by the name "gamelogs" already exist, data from gl2012 will be appended to the table. If append is set to FALSE, the table "gamelogs" (if it exists) will be overwritten.

- The `field.types` argument supplies a named vector of data types for
  the MySQL columns. If this argument is left blank, `dbWriteTable()` will
  attempt to guess the optimal values. In this case, we choose to specify
  values for certain variables because we want to enforce uniformity across
  various years.

```
if (dbExistsTable(con, "gamelogs")) {
  dbRemoveTable(con, "gamelogs")
}
con |>
  dbWriteTable(
    name = "gamelogs", value = gl2012,
    append = FALSE,
    field.types = c(
      CompletionInfo = "varchar(50)",
      AdditionalInfo = "varchar(255)",
      HomeBatting1Name = "varchar(50)",
      HomeBatting2Name = "varchar(50)",
      HomeBatting3Name = "varchar(50)",
      HomeBatting4Name = "varchar(50)",
      HomeBatting5Name = "varchar(50)",
      HomeBatting6Name = "varchar(50)",
      HomeBatting7Name = "varchar(50)",
      HomeBatting8Name = "varchar(50)",
      HomeBatting9Name = "varchar(50)",
      HomeManagerName = "varchar(50)",
      VisitorStartingPitcherName = "varchar(50)",
      VisitorBatting1Name = "varchar(50)",
      VisitorBatting2Name = "varchar(50)",
      VisitorBatting3Name = "varchar(50)",
      VisitorBatting4Name = "varchar(50)",
      VisitorBatting5Name = "varchar(50)",
      VisitorBatting6Name = "varchar(50)",
      VisitorBatting7Name = "varchar(50)",
      VisitorBatting8Name = "varchar(50)",
      VisitorBatting9Name = "varchar(50)",
      VisitorManagerName = "varchar(50)",
      HomeLineScore = "varchar(30)",
      VisitorLineScore = "varchar(30)",
      SavingPitcherName = "varchar(50)",
      ForfeitInfo = "varchar(10)",
      ProtestInfo = "varchar(10)",
      UmpireLFID = "varchar(8)",
      UmpireRFID = "varchar(8)",
      UmpireLFName = "varchar(50)",
      UmpireRFName = "varchar(50)"
    )
  )
```

To verify that the data now resides in the MySQL Server, we can query it using **dplyr**.

```
gamelogs <- con |>
  tbl("gamelogs")
head(gamelogs)
```

```
# Source:    SQL [6 x 161]
# Database: mysql  [abdwr@localhost:NA/abdwr]
     Date DoubleHeader DayOfWeek VisitingTeam VisitingTeamLeague
    <dbl>        <dbl> <chr>     <chr>        <chr>
1  2.01e7            0 Wed       SEA          AL
2  2.01e7            0 Thu       SEA          AL
3  2.01e7            0 Wed       SLN          NL
4  2.01e7            0 Thu       TOR          AL
5  2.01e7            0 Thu       BOS          AL
6  2.01e7            0 Thu       WAS          NL
# i 156 more variables: VisitingTeamGameNumber <dbl>,
#   HomeTeam <chr>, HomeTeamLeague <chr>,
#   HomeTeamGameNumber <dbl>, VisitorRunsScored <dbl>,
#   HomeRunsScore <dbl>, LengthInOuts <dbl>, DayNight <chr>,
#   CompletionInfo <chr>, ForfeitInfo <chr>, ProtestInfo <chr>,
#   ParkID <chr>, Attendance <dbl>, Duration <dbl>,
#   VisitorLineScore <chr>, HomeLineScore <chr>, ...
```

In Section 11.4.1, we provided code for appending one season of game logs into a MySQL table. However, we have demonstrated in previous chapters that it is straightforward to use R to work with a single season of game logs. To fully appreciate the advantages of storing data in a RDBMS, we will populate a MySQL table with game logs going back through baseball history. With a historical database and an R connection, we demonstrate the use of R to perform analysis over multiple seasons.

We write a simple function `append_game_logs()` that combines the previous two steps of importing the game log data into R, and then transfers those data to a MySQL table.[4] The whole process may take several minutes. If one is not interested in downloading files dating back to 1871, seasons from 1995 are sufficient for reproducing the example of the next section.

The function `append_game_logs()` takes the following parameters as inputs:

- `conn` is a **DBI** connection to a database.

---

[4]The downloading of data from Retrosheet is performed by the previously presented `retrosheet_gamelog()` function; thus the reader has to make sure said function is loaded for the code in this section to work.

- **season** indicates the season one wants to download from Retrosheet and append to the MySQL database. By default the function will work on seasons from 1871 to 2022.

```
append_game_logs <- function(conn, season) {
  message(paste("Working on", season, "season..."))
  one_season <- retrosheet_gamelog(season)
  conn |>
    dbWriteTable(
      name = "gamelogs", value = one_season, append = TRUE
    )
}
```

Next, we remove the previous games using the TRUNCATE TABLE SQL command and then fill the table by iterating the process of append_game_logs using map().

```
dbSendQuery(con, "TRUNCATE TABLE gamelogs;")
map(1995:2017, append_game_logs, conn = con)
```

Now we have many years worth of game logs.

```
gamelogs |>
  group_by(year = str_sub(Date, 1, 4)) |>
  summarize(num_games = n())
```

```
# Source:    SQL [?? x 2]
# Database: mysql  [abdwr@localhost:NA/abdwr]
   year  num_games
   <chr>    <int64>
 1 1995        2017
 2 1996        2267
 3 1997        2266
 4 1998        2432
 5 1999        2428
 6 2000        2429
 7 2001        2429
 8 2002        2426
 9 2003        2430
10 2004        2428
# i more rows
```

11.5   Querying Data from R

11.5.1   Retrieving data from SQL

All **DBI** backends support dbGetQuery(), which retrieves the results of an SQL query from the database. As the purpose of having data stored in a MySQL database is to selectively import data into R for particular analysis, one typically selectively reads data into R by querying one or more tables of the database.

For example, suppose one is interested in comparing the attendance of the two Chicago teams by day of the week since the 2006 season. The following code retrieves the raw data in R.

```
query <- "
SELECT date, hometeam, dayofweek, attendance
FROM gamelogs
WHERE Date > 20060101
  AND HomeTeam IN ('CHN', 'CHA');
"
chi_attendance <- dbGetQuery(con, query)
slice_head(chi_attendance, n = 6)
```

	date	hometeam	dayofweek	attendance
1	20060402	CHA	Sun	38802
2	20060404	CHA	Tue	37591
3	20060405	CHA	Wed	33586
4	20060407	CHN	Fri	40869
5	20060408	CHN	Sat	40182
6	20060409	CHN	Sun	39839

The dbGetQuery() function queries the database. Its arguments are the connection handle established previously (conn) and a string consisting of a valid SQL statement (query). Readers familiar with SQL will have no problem understanding the meaning of the query. For those unfamiliar with SQL, we present here a brief explanation of the query, inviting anyone who is interested in learning about the language to look for the numerous resources devoted to the subject (see for example, https://dev.mysql.com/doc/refman/8.2/en/select.html).

The first row in the SQL statement indicates the columns of the table that are to be *select*-ed (in this case date, hometeam, dayofweek, and attendance). The second line states *from* which table they have to be retrieved (gamelogs). Finally, the *where* clause specifies conditions for the rows that are to be retrieved: the date has to be greater than 20060101 *and* the value of hometeam has to be either CHN or CHA.

Alternatively, one could use the **dplyr** interface to MySQL through the gamelogs object we created earlier.

```
gamelogs |>
  filter(Date > 20060101, HomeTeam %in% c('CHN', 'CHA')) |>
  select(Date, HomeTeam, DayOfWeek, Attendance) |>
  head()
```

```
# Source:    SQL [6 x 4]
# Database: mysql  [abdwr@localhost:NA/abdwr]
      Date HomeTeam DayOfWeek Attendance
     <dbl> <chr>    <chr>          <dbl>
1 20060402 CHA      Sun            38802
2 20060404 CHA      Tue            37591
3 20060405 CHA      Wed            33586
4 20060407 CHN      Fri            40869
5 20060408 CHN      Sat            40182
6 20060409 CHN      Sun            39839
```

dplyr even contains a function called show_query() that will translate your **dplyr** pipeline into a valid SQL query. Note the similarities and differences between the SQL code we wrote above and the translated SQL code below.

```
gamelogs |>
  filter(Date > 20060101, HomeTeam %in% c('CHN', 'CHA')) |>
  select(Date, HomeTeam, DayOfWeek, Attendance) |>
  show_query()
```

```
<SQL>
SELECT `Date`, `HomeTeam`, `DayOfWeek`, `Attendance`
FROM `gamelogs`
WHERE (`Date` > 20060101.0) AND (`HomeTeam` IN ('CHN', 'CHA'))
```

11.5.2   Data cleaning

Before we can plot these data, we need to clean up two things. We first use the ymd() function from the **lubridate** package to convert the number that encodes the date into a Date field in R. Next, we set the attendance of the games reporting zero to NA using the na_if() function.[5]

---

[5]In case of single admission doubleheaders (i.e., when two games are played on the same day and a single ticket is required for attending both) the attendance is reported only for the second game, while it is set at zero for the first.

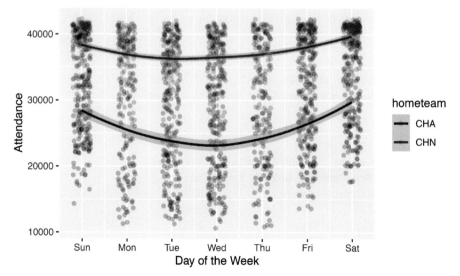

FIGURE 11.1
Comparison of attendance by day of the week on games played at home by
the Cubs (CHN} and the White Sox (CHA).

```
chi_attendance <- chi_attendance |>
  mutate(
    the_date = ymd(date),
    attendance = na_if(attendance, 0)
  )
```

We show a graphical comparison between attendance at the two Chicago
ballparks in Figure 11.1. In order to get `geom_smooth()` to work, the variable
on the horizontal axis must be numeric. Thus, we use the `wday()` function
from **lubridate** to compute the day of the week (as a number) from the date.
Getting the axis labels to display as abbreviations requires using `wday()` again,
but with the `label` argument set to `TRUE`.

```
ggplot(
  chi_attendance,
  aes(
    x = wday(the_date), y = attendance,
    color = hometeam
  )
) +
  geom_jitter(height = 0, width = 0.2, alpha = 0.2) +
  geom_smooth() +
  scale_y_continuous("Attendance") +
```

```
scale_x_continuous(
  "Day of the Week", breaks = 1:7,
  labels = wday(1:7, label = TRUE)
) +
scale_color_manual(values = crc_fc)
```

We note that the Cubs typically draw more fans than the White Sox, while both teams see larger crowds on weekends.

11.5.3 Coors Field and run scoring

As an example of accessing multiple years of data, we explore the effect of Coors Field (home of the Colorado Rockies in Denver on run scoring through the years. Coors Field is a peculiar ballpark because it is located at an altitude of about one mile above sea level. The air is thinner than in other stadiums, resulting in batted balls that travel farther and curveballs that are "flatter". Lopez, Matthews, and Baumer (2018) estimate that Coors Field confers the largest home advantage in all of baseball.

We retrieve data for the games played by the Rockies—either at home or on the road—since 1995 (the year they moved to Coors Field) using a SQL query and the function dbGetQuery()[6].

```
query <- "
SELECT date, parkid, visitingteam, hometeam,
  visitorrunsscored AS awR, homerunsscore AS hmR
FROM gamelogs
WHERE (HomeTeam = 'COL' OR VisitingTeam = 'COL')
  AND Date > 19950000;
"
rockies_games <- dbGetQuery(con, query)
```

The game data is conveniently stored in the `rockies_games` data frame. We compute the sum of runs scored in each game by adding the runs scored by the home team and the visiting team. We also add a new column `coors` indicating whether the game was played at Coors Field.[7]

```
rockies_games <- rockies_games |>
  mutate(
    runs = awR + hmR,
```

---

[6]The keyword AS in SQL has the purpose of assigning different names to columns. Thus visitorrunsscored AS awR tells SQL that, in the results returned by the query, the column visitorrunsscored will be named awR.

[7]Retrosheet code for Coors Field is DEN02. A list of all ballpark codes is available at https://www.retrosheet.org/parkcode.txt.

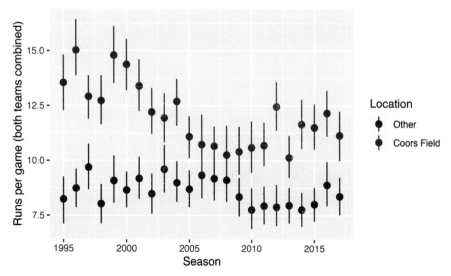

FIGURE 11.2
Comparison of runs scored by the Rockies and their opponents at Coors Field
and in other ballparks.

```
  coors = parkid == "DEN02"
)
```

We compare the offensive output by the Rockies and their opponents at Coors
and other ballparks graphically in Figure 11.2.

```
ggplot(
  rockies_games,
  aes(x = year(ymd(date)), y = runs, color = coors)
) +
  stat_summary(fun.data = "mean_cl_boot") +
  xlab("Season") +
  ylab("Runs per game (both teams combined)") +
  scale_color_manual(
    name = "Location", values = crc_fc,
    labels = c("Other", "Coors Field")
  )
```

We use the `stats_summary()` layer to summarize the y values at every unique
value of x. The `fun.data` argument lets the user specify a summarizing function; in this case `mean_cl_boot()` implements a nonparametric bootstrap
procedure for obtaining confidence bands for the population mean. The output
resulting from this layer are the vertical bars appearing for each data point.

The `scale_linetype_discrete()` layer is used for labeling the series (`name` argument) and assigning a name to the legend (`labels`).

From Figure 11.2 one notices how Coors Field has been an offense-friendly park, boosting run scoring by as much as six runs per game over the course of a season. However, the effect of the Colorado ballpark has somewhat decreased in the new millennium, displaying smaller differences in the 2006–2011 period. One reason for Coors becoming less of an extreme park is the installation of a humidor. Since the 2002 season, baseballs have been stored in a room at a higher humidity prior to each game, with the intent of compensating for the unusual natural atmospheric conditions.[8]

11.6 Building Your Own Baseball Database

Section 11.4.1 illustrated populating a MySQL database from within R by creating a table of Retrosheet game logs. Several so-called *SQL dumps* are available online for creating and filling databases with baseball data. SQL dumps are simple text files (featuring a `.sql` extension) containing SQL instructions for creating and filling SQL tables.

11.6.1 Lahman's database

Sean Lahman provides his historical database of seasonal stats in several formats. There is also the **Lahman** package for R, which makes these data available in memory. However, the database is also available as a SQL dump, which can be downloaded from http://seanlahman.com/download-baseball-database/ (look for the *2019 - MySQL version*). Unfortunately, the MySQL dump files are no longer supported, but older versions are still available.

There are several options for importing this file into SQL. It is beyond the scope of this book to illustrate all said processes; thus here we provide a command to obtain the desired result from a terminal. In order for the following code to work, the MySQL service should be running[9] (see Section 11.2).

```
mysql -u username -p lahmansbaseballdb < lahman-mysql-dump.sql
```

Note that the SQL dump file creates a new database named `lahmansbaseballdb`.

---

[8] For a detailed analysis of the humidor's effects, see Nathan (2011).

[9] Also, make sure to change the directory containing the `.sql` file and the user name and the password as appropriate.

11.6.2   Retrosheet database

In Appendix A we provide R code to download Retrosheet files and transform them in formats easily readable by R. By slightly adapting the code provided in Section 11.4.2, one can also append them to a MySQL database.

There are many software packages available on the Internet to process Retrosheet files. We built our database using the **baseballr** package.

11.6.3   Statcast database

As illustrated in Appendix C, the **baseballr** package provides the `statcast_search()` function for downloading Statcast data from Baseball Savant. In Chapter 12, we show how the **abdwr3edata** package can leverage this functionality to download multiple seasons of Statcast data. The chapter also discusses various alternative data storage options and their respective strengths and weaknesses.

11.7   Calculating Basic Park Factors

Park factors (usually abbreviated as PF) have been used for decades by baseball analysts as a tool for contextualizing the effect of the ballpark when assessing the value of players. Park factors have been calculated in several ways and in this section we illustrate a very basic approach, focusing on year 1996, one of the most extreme seasons for Coors Field, as displayed in Figure 11.2.

In the explanations that will follow, we presume the reader has Retrosheet data for the 1990s in her database.[10] The following code will set up a suitable database using the **baseballr** package.

```
library(baseballr)
retro_data <- baseballr::retrosheet_data(
  here::here("data_large/retrosheet"),
  1990:1999
)
events <- retro_data |>
  map(pluck, "events") |>
  bind_rows() |>
  as_tibble()

con |>
```

---

[10]Refer to Section 11.6.2 for performing the necessary steps to get the data into a MySQL database.

```
    dbWriteTable(name = "events", value = events)

  events_db <- con |>
    tbl("events")
```

Once this process is complete, the database should contain a table called `events` with more than 1.7 million rows. Note that while storing the full table in R as the `tibble` object `events` takes up nearly 1 GiB of memory, pushing that data to the database and using the **dplyr** interface to the database as the object `events_db` occupies almost no space in memory (of course, it still takes up 1 GiB on disk).

11.7.1 Loading the data into R

As a first step, we connect to the MySQL database and retrieve the desired data. Using a SQL query, we *select* the columns containing the home and away teams and the event code *from* the `events` table, keeping only the rows *where* the year is 1996 *and* the event code corresponds to one indicating a batted ball (see Appendix A). The results of the query are stored in the `hr_PF` data frame in R.

```
  query <- "
  SELECT away_team_id, LEFT(game_id, 3) AS home_team_id, event_cd
  FROM events
  WHERE year = 1996
    AND event_cd IN (2, 18, 19, 20, 21, 22, 23);
  "
  hr_PF <- dbGetQuery(con, query)
  dim(hr_PF)
```

```
[1] 130437        3
```

11.7.2 Home run park factor

A ballpark can have different effects on various player performance statistics. The unique configuration of Fenway Park in Boston, for example, enhances the likelihood of a batted ball to become a double, especially flyballs to left, which often carom off the Green Monster. On the other hand, home runs are rare on the right side of Fenway Park due to the unusually deep right-field fence.

In this example we explore the stadium effect on home runs in 1996, by calculating *park factors* for home runs. To begin, we create a new column `was_hr`, which indicates the occurrence of a home run for every row in the `hr_PF` data frame.

```
hr_PF <- hr_PF |>
  mutate(was_hr = ifelse(event_cd == 23, 1, 0))
```

Next, we compute the frequency of home runs per batted ball for all MLB teams both at home and on the road. Note that this requires two passes through the entire data frame.

```
ev_away <- hr_PF |>
  group_by(team_id = away_team_id) |>
  summarize(hr_event = mean(was_hr)) |>
  mutate(type = "away")

ev_home <- hr_PF |>
  group_by(team_id = home_team_id) |>
  summarize(hr_event = mean(was_hr)) |>
  mutate(type = "home")
```

We then combine the two resulting data frames and use the `pivot_wider()` function to put the home and away home run frequencies side-by-side.

```
ev_compare <- ev_away |>
  bind_rows(ev_home) |>
  pivot_wider(names_from = type, values_from = hr_event)

ev_compare
```

```
# A tibble: 28 x 3
   team_id   away    home
   <chr>    <dbl>   <dbl>
 1 ATL      0.0323  0.0372
 2 BAL      0.0488  0.0477
 3 BOS      0.0385  0.0443
 4 CAL      0.0387  0.0483
 5 CHA      0.0424  0.0349
 6 CHN      0.0374  0.0407
 7 CIN      0.0403  0.0393
 8 CLE      0.0440  0.0372
 9 COL      0.0341  0.0538
10 DET      0.0457  0.0506
# i 18 more rows
```

Park factors are typically calculated so that the value 100 indicates a neutral ballpark (one that has no effect on the particular statistic) while values over 100 indicate playing fields that increase the likelihood of the event (home run in this case) and values under 100 indicate ballparks that decrease the likelihood of the event.

We compute the 1996 home run park factors with the following code, and use `arrange()` to display the ballparks with the largest and smallest park factors.

```
ev_compare <- ev_compare |>
  mutate(pf = 100 * home / away)
ev_compare |>
  arrange(desc(pf)) |>
  slice_head(n = 6)
```

```
# A tibble: 6 x 4
  team_id   away   home     pf
  <chr>    <dbl>  <dbl>  <dbl>
1 COL     0.0341 0.0538   158.
2 CAL     0.0387 0.0483   125.
3 ATL     0.0323 0.0372   115.
4 BOS     0.0385 0.0443   115.
5 DET     0.0457 0.0506   111.
6 SDN     0.0294 0.0320   109.
```

Coors Field is at the top of the HR-friendly list, displaying an extreme value of 158—this park boosted home run frequency by over 50% in 1996!

```
ev_compare |>
  arrange(pf) |>
  slice_head(n = 6)
```

```
# A tibble: 6 x 4
  team_id   away   home    pf
  <chr>    <dbl>  <dbl> <dbl>
1 LAN     0.0360 0.0256  71.2
2 HOU     0.0344 0.0272  79.1
3 NYN     0.0363 0.0289  79.5
4 CHA     0.0424 0.0349  82.2
5 CLE     0.0440 0.0372  84.6
6 FLO     0.0316 0.0271  85.7
```

At the other end of the spectrum was Dodger Stadium in Los Angeles, featuring a home run park factor of 71, meaning that it suppressed home runs by nearly 30% relative to the league average park.

11.7.3 Assumptions of the proposed approach

The proposed approach to calculating park factors makes several simplifying assumptions. The first assumption is that the home team always plays at the same home ballpark. While that is true for most teams in most seasons, sometimes alternate ballparks have been used for particular games. For example, during the 1996 season, the Oakland Athletics played their first 16 home games in Cashman Field (Las Vegas, NV) while renovations at the Oakland-Alameda County Coliseum were being completed. In the same year, as a marketing move by MLB, the San Diego Padres had a series of three games against the New York Mets at Estadio de Béisbol Monterrey in Mexico.[11]

Another assumption of the proposed approach is that a single park factor is appropriate for all players, without considering how ballparks might affect some categories of players differently. Asymmetric outfield configurations, in fact, cause playing fields to have unequal effects on right-handed and left-handed players. For example, the aforementioned Green Monster in Boston, being situated in left field, comes into play more frequently when right-handed batters are at the plate; and the latest Yankee Stadium has seen left-handed batters take advantage of the short distance of the right field fence.

Finally, the proposed park factors (as well as most published versions of park factors) essentially ignore the players involved in each event (in this case the batter and the pitcher). As teams rely more on the analysis of play-by-play data, they typically adapt their strategies to accommodate the peculiarities of ballparks. For example, while the diminished effect of Coors Field on run scoring displayed in Figure 11.2 is mostly attributable to the humidor, part of the effect is certainly due to teams employing different strategies when playing in this park. For example, teams can use pitchers who induce a high number of groundballs that may be less affected by the rarefied air.

11.7.4   Applying park factors

In the 1996 season, four Rockies players hit 30 or more home runs: Andrés Galarraga led the team with 47, followed by Vinnie Castilla and Ellis Burks (tied at 40) and Dante Bichette (31). Behind them Larry Walker had just 18 home runs, but in very limited playing time due to injuries. In fact, Walker's HR/AB ratio was second only to Galarraga's. Their offensive output was certainly boosted by playing 81 of their games in Coors Field. Using the previously calculated park factors, one can estimate the number of home runs Galarraga would have hit in a neutral park environment.

We begin by retrieving from the MySQL database every one of Galarraga's 1996 plate appearances ending with a batted ball, and defining a column `was_hr` is defined indicating whether the event was a home run, as was done in Section 11.7.2).

```
query <- "
SELECT away_team_id, LEFT(game_id, 3) AS home_team_id, event_cd
```

---

[11]A list of games played in alternate sites is displayed on the Retrosheet website at the url https://www.retrosheet.org/neutral.htm.

```
FROM events
WHERE year = 1996
  AND event_cd IN (2, 18, 19, 20, 21, 22, 23)
  AND bat_id = 'galaa001';
"
andres <- dbGetQuery(con, query) |>
  mutate(was_hr = ifelse(event_cd == 23, 1, 0))
```

We add the previously calculated park factors to the `andres` data frame. This is done by merging the data frames `andres` and `ev_compare` using the `inner_join()` function with the columns `home_team_id` and `team_id` as keys. In the merged data frame `andres_pf`, we calculate the mean park factor for Galarraga's plate appearances using `summarize()`.

```
andres_pf <- andres |>
  inner_join(ev_compare, by = c("home_team_id" = "team_id")) |>
  summarize(mean_pf = mean(pf))
andres_pf
```

```
  mean_pf
1     129
```

The composite park factor for Galarraga, derived from the 252 batted balls he had at home and the 225 he had on the road (ranging from 9 in Dodger Stadium in Los Angeles to 23 at the Astrodome in Houston), indicate Andrés had his home run frequency increased by an estimated 29% relative to a neutral environment. In order to get the estimate of home runs in a neutral environment, we divide Galarraga's home runs by his average home run park factor divided by 100.

```
47 / (andres_pf / 100)
```

```
  mean_pf
1    36.4
```

According to our estimates, Galarraga's benefit from the ballparks he played in (particularly his home Coors Field) amounted to roughly $47 - 36 = 11$ home runs in the 1996 season.

11.8 Further Reading

Chapter 2 of Adler (2006) has detailed instructions on how to obtain and install MySQL, and on how to set up an historical baseball database with

Retrosheet data. Hack #56 (Chapter 5 of the same book) provides SQL code for computing and applying Park Factors. Section F.2 of Benjamin S. Baumer, Kaplan, and Horton (2021b) contains step-by-step instructions more complete than those presented in Section 11.2.

MySQL reference manuals are available in several formats at http://dev.my sql.com/doc/ on the MySQL website. The HTML online version features a search box, which allows users to quickly retrieve pages pertaining to specific functions.

11.9 Exercises

1. Runs Scored at the Astrodome

a. Using the `dbGetQuery()` function from the **DBI** package, select games featuring the Astros (as either the home or visiting team) during the years when the Astrodome was their home park (i.e., from 1965 to 1999).

b. Draw a plot to visually compare through the years the runs scored (both teams combined) in games played at the Astrodome and in other ballparks.

2. Astrodome Home Run Park Factor

a. Select data from one season between 1965 and 1999. Keep the columns indicating the visiting team identifier, the home team identifier and the event code, and the rows identifying ball-in-play events. Create a new column that identifies whether a home run has occurred.

b. Prepare a data frame containing the team identifier in the first column, the frequency of home runs per batted ball when the team plays on the road in the second column, and the same frequency when the team plays at home in the third column.

c. Compute home run park factors for all MLB teams and check how the domed stadium in Houston affected home run hitting.

3. Applying Park Factors to "Adjust" Numbers

a. Using the same season selected for the previous exercise, obtain data from plate appearances (ending with the ball being hit into play) featuring one Astros player of choice. The exercise can either be performed on plate appearances featuring an Astros pitcher on the mound or an Astros batter at the plate. For example, if the selected season is 1988, one might be interested in discovering how the Astrodome affected the number of home runs surrendered by veteran pitcher Nolan Ryan (Retrosheet id: `ryann001`) or the number of home runs hit by rookie catcher Craig Biggio (id: `biggc001`).

b. As shown in Section 11.7.4, merge the selected player's data with the Park Factors previously calculated and compute the player's individual Park Factor (which is affected by the different playing time the player had in the various ballparks) and use it to estimate a "fair" number of home runs hit (or surrendered if a pitcher was chosen).

4. Park Factors for Other Events

a. Park Factors can be estimated for events other than Home Runs. The SeamHeads.com Ballpark Database, for example, features Park Factors for seven different events, plus it offers split factors according to batters' handedness. See for example the page for the Astrodome: http://www.se amheads.com/ballparks/ballpark.php?parkID=HOU02&tab=pf1.

b. Choose an event (even different from the seven shown at SeamHeads) and calculate how ballparks affect its frequency. As a suggestion, the reader may want to look at seasons in the '80s, when artificial turf was installed in close to 40% of MLB fields, and verify whether parks with concrete/synthetic grass surfaces featured a higher frequency of batted balls (home runs excluded) converted into outs.[12]

5. Length of Game

Major League Baseball instituted a pitch clock for the 2023 season, as one of several measures to reduce the length of games. The Retrosheet game logs contain a variable `Duration` that measures the length of each game, in minutes. Use the `retrosheet_gamelog()` function to compare the length of games from 2022 to 2023. Draw a box plot to illustrate the distribution of length of game.

6. Length of Game (continued)

Extend your analysis from the previous question to go back as far as is necessary to find the most recent year (prior to 2023) that had an average length of game less than that of 2023.

---

[12]SeamHeads provides information on the surface of play in each stadium's page. For example, in the previously mentioned page relative to the Astrodome, if one hovers the mouse over the ballpark name in a given season, a pop-up will appear providing information both on the ballpark cover and its playing field surface. SeamHeads is currently providing its ballpark database as a zip archive containing comma-separated-value (.csv) files that can be easily read by R: the link for downloading it is found at the bottom of each page in the ballpark database section.

12

Working with Large Data

12.1 Introduction

In Chapter 11, we set up and populated a MySQL database using the **RMariaDB** package for R. That database contained game logs and play-by-play data gathered from Retrosheet. In this chapter, we expand on this work by building databases of pitch-by-pitch Statcast data downloaded from Baseball Savant. Along the way, in addition to MySQL we explore several different ways to store and represent a dataset that spans four years of play (2020–2023). Finally, we compare and contrast the strengths of weaknesses of these various approaches and take these databases for a few test drives.

In Section C.10, we show how to use the `statcast_search()` function from the **baseballr** package to download Statcast data from Baseball Savant. Because there is a row of data for each pitch thrown in Major League Baseball, these data can quickly become large enough to complicate our workflow. For the most part, the data that we have used in this book has been *small*, in the sense that it only comprises a few *kilobytes* or *megabytes*, which is far less than the amount of physical memory (i.e. *random access memory*) in most personal computers. Even the Retrosheet data, which covers every play, is less than 100 MB for a full season. However, as we will see below, the Statcast data is an *order of magnitude* larger, and will occupy several hundred megabytes per season, and several gigabytes over multiple seasons. Working with data of this magnitude can impede or overwhelm a personal computer if one is not careful. As such, in this chapter we introduce new tools for acquiring multiple years worth of Statcast, and then storing that data efficiently. Mercifully, regardless of what data storage format we choose, the magic of **dplyr** and **dbplyr** makes the process of analyzing the data the same.

12.2 Acquiring a Year's Worth of Statcast Data

To further explore many sabermetric ideas, it's useful to collect the in-play data for all plays in a particular season. One can acquire the relevant data via the `statcast_search()` function provided by **baseballr**. However, because this function only returns at most 25,000 observations, and there were more than 700,000 pitches thrown in 2023, compiling the full season's worth of data takes some effort. To that end we show how one can iterate the process of downloading Statcast data on a daily basis, and then combine that data into the full-season data sets that we use in this book.

On a typical day during the regular season, around 4500 pitches are thrown. This means that in a typical week, more than 30,000 pitches are thrown, which exceeds the 25,000 limit returned by `statcast_search()`. This means that we can't safely download data on a weekly basis. As a workaround, we write the following function to download Statcast data for a single day and write that data to an appropriately-named CSV in a specified directory `dir`. The code for the `statcast_daily()` function is included in the **abdwr3edata** package, and reproduced here.

```
abdwr3edata::statcast_daily
```

```
function(the_date = lubridate::now(), dir = getwd()) {
  if (!dir.exists(dir)) {
    dir.create(dir, recursive = TRUE)
  }

  filename <- paste0("sc_", lubridate::as_date(the_date), ".csv")
  file_path <- fs::path(dir, filename)

  # if the file already exists, read it.
  if (file.exists(file_path)) {
    x <- file_path |>
      readr::read_csv() |>
      suppressMessages()
    if (nrow(x) > 0) {
      message(
        paste(
          "Found", nrow(x), "observations in", file_path, "..."
        )
      )
    }
    return(NULL)
  }
```

```
  # the file doesn't exist or doesn't have data, get it
  message(paste("Retrieving data from", the_date))
  x <- baseballr::statcast_search(
    start_date = lubridate::as_date(the_date),
    end_date = lubridate::as_date(the_date),
    player_type = "batter"
  ) |>
    dplyr::filter(game_type == "R")

  if (nrow(x) > 0) {
    message(paste("Writing", file_path, "..."))
    x |>
      readr::write_csv(file = fs::path(dir, filename))
  }
  return(NULL)
}
<bytecode: 0x582ac35576e0>
<environment: namespace:abdwr3edata>
```

In this case, we don't want to store the data in the working directory. We want to store it in a directory named `statcast_csv`, and we need to create that directory if it doesn't exist already. Using the `path()` function from the **fs** package ensures that the file paths we create will be valid on any operating system.

```
  library(fs)
  data_dir <- here::here("data_large")
  statcast_dir <- path(data_dir, "statcast_csv")
  if (!dir.exists(statcast_dir)) {
    dir.create(statcast_dir)
  }
```

Now, we create a vector of dates for which we want to download Statcast data. For example, regular season games during the 2023 season were played between March 30 and November 6. These were the 89th and 274th days of the year, respectively. To extend `statcast_daily()` to a full season, we'll have to iterate the function over a series of dates. We can use the `parse_date_time()` function from the **lubridate** package to convert a vector of integers to a vector of dates.

```
  mlb_2023_dates <- 89:274 |>
    parse_date_time("%j") |>
    as_date()
  head(mlb_2023_dates)
```

```
[1] "2024-03-29" "2024-03-30" "2024-03-31" "2024-04-01"
[5] "2024-04-02" "2024-04-03"
```

The `walk()` function from the **purrr** package will allow us to successively apply
the `statcast_daily()` function to each of these 186 days. `walk()` is similar to
`map()` but doesn't return anything. Since `statcast_daily()` always returns
NULL, the effect is the same. Note that since `statcast_daily()` writes a CSV
file for each day, you can safely run this function over and over again without
duplicating any work. The code for the `statcast_season()` function shown
below pulls these steps together. Note that if we want to be able to use this
function over multiple years, we need to be a little bit more conservative with
our choice of beginning and ending dates. While these dates vary from one
year to another, they will typically be close to the first of April and the first of
November.

```
    abdwr3edata::statcast_season
```

```
function(
    year = lubridate::year(lubridate::now()), dir = getwd()
) {
  if (!dir.exists(dir)) {
    dir.create(dir, recursive = TRUE)
  }

  mlb_days <- 80:280
  mlb_dates <- mlb_days |>
    paste(year) |>
    lubridate::parse_date_time("%j %Y") |>
    lubridate::as_date()

  mlb_dates |>
    purrr::walk(statcast_daily, dir)
}
<bytecode: 0x582ac71bac90>
<environment: namespace:abdwr3edata>
```

Once the directory is populated with daily CSV files, we can read them all
into one large data frame with a call to `read_csv()`. This functionality is
encapsulated in the `statcast_read_csv()` function.

```
    abdwr3edata::statcast_read_csv
```

```
function(dir = getwd(), pattern = "*.csv") {
  dir |>
    list.files(pattern = pattern, full.names = TRUE) |>
```

```
  readr::read_csv() |>
  dplyr::bind_rows()
}
<bytecode: 0x582ac739aac0>
<environment: namespace:abdwr3edata>
```

Putting it all together, we run the `statcast_season()` function to verify that we've downloaded all of the data for the 2023 season, and then once that is complete, we run the `statcast_read_csv()` function to read the various CSV files we've downloaded.

```
library(abdwr3edata)

# skip this step while building the book!
# statcast_season(2023, dir = statcast_dir)

sc2023 <- statcast_dir |>
  statcast_read_csv(pattern = "2023.+\\.csv")
```

To check the validity of the data, we can spot check certain statistics against their known values. First, if 30 teams play 162 games against each other, we should see 2430 total games. Second, according to Baseball-Reference, there were 5,868 home runs hit in 2023. How many do we see in our data?

```
sc2023 |>
  group_by(game_type) |>
  summarize(
    num_games = n_distinct(game_pk),
    num_pitches = n(),
    num_hr = sum(events == "home_run", na.rm = TRUE)
  )
```

```
# A tibble: 1 x 4
  game_type num_games num_pitches num_hr
  <chr>         <int>       <int>  <int>
1 R              2430      717945   5868
```

Our data appears to be accurate for the full season.

If we want to acquire *multiple* years of Statcast data, we simply iterate the `statcast_season()` function over multiple years.

In the remainder of this chapter, we analyze data across the 2020–2023 seasons.

12.3   Storing Large Data Efficiently

Once we're satisfied with the integrity of our data, we should store the data in a format that makes it easy to work with in the future. Downloading the data for each day took a long time, and there is no reason to keep pinging the Baseball Savant servers once we already have a copy of the data.

As noted above, a full season of Statcast data contains over 700,000 rows and nearly 100 variables. This takes up about half of a *gigabyte* of space in memory in R.

```
sc2023 |>
  object.size() |>
  print(units = "MB")
```

500.7 Mb

While most computers now have at least 16 GB of memory, working with data of this magnitude (especially if you want to work with data across multiple seasons) can quickly become burdensome. We can help ourselves by writing the data to disk and storing it in an efficient format.

We saw above that the data frame containing a full season's worth of Statcast data occupies about half a gigabtye in memory. We built that data frame by combining many small CSVs, each containing one day's worth of data. How much space do those CSV files occupy *on disk*?

```
statcast_dir |>
  list.files(full.names = TRUE) |>
  str_subset("2023") |>
  file.size() |>
  sum() / 1024^2
```

[1] 373

In this case, the CSVs occupy about 75% as much space on disk as the data do when read into memory. That's fine—but we can do better!

The main advantage of CSVs is that they are human-readable and editable, which makes them easy to understand, and because they are so commonly-used, they can be read and written by virtually any program designed for working with data. However, CSVs are not a very space-efficient file format. Let's explore some different options.

12.3.1   Using R's internal data format

Suppose now that we want to work with multiple years of Statcast data. For example, suppose we want to investigate pandemic-era trends using data from the 2020–2023 seasons. Should we store these data in four files or one? If we keep the data in four files (one for each year), then analyzing it across multiple years will be cumbersome, since we would have to read the data into R separately, and then combine the resulting data frames into one large data frame. That large data frame will occupy multiple gigabytes of *memory*, which could make things sluggish. Alternatively, we could do this once, and then write one big *.rds file that contained all four years worth of data. But then *that* file would be large, and we'd have to redo that process every time new data came in.

Once the data has been read into R, we can use the `write_rds()` function to write the full-season data frame to disk for safekeeping. This uses R's internal binary data storage format, which is much more space-efficient than CSVs. However, the .rds format is designed to work with R, so while it's great if you're going to be working exclusively in R, it isn't that useful if you want to share your data with users of other applications (like Python).

The `statcast_write_rds()` function wraps the `write_rds()` function, but first splits the data into groups based on the year. It will write a different, appropriately-named *.rds file for each year present in the data frame that you give it.

```
abdwr3edata::statcast_write_rds
```

```
function(x, dir = getwd(), ...) {
  tmp <- x |>
    dplyr::group_by(year = lubridate::year(game_date))
  years <- tmp |>
    dplyr::group_keys() |>
    dplyr::pull(year)
  tmp |>
    dplyr::group_split() |>
    rlang::set_names(years) |>
    purrr::map(
      ~readr::write_rds(
        .x,
        file = fs::path(
          dir,
          paste0(
            "statcast_",
            max(lubridate::year(dplyr::pull(.x, game_date))),
            ".rds"
```

```
          )
        ),
        compress = "xz",
#          ...
      )
    )
  list.files(dir, pattern = "*.rds", full.names = TRUE)
}
<bytecode: 0x582ac79e6c38>
<environment: namespace:abdwr3edata>
```

In this case, the directory of CSV files takes up nearly 6 times more space (373 Mb for the 2023 files) than the single .rds file that contains the exact same data (64 Mb)!

```
disk_space_rds <- data_dir |>
  path("statcast_rds") |>
  dir_info(regexp = "*.rds") |>
  select(path, size) |>
  mutate(
    path = path_file(path),
    format = "rds"
  )
disk_space_rds
```

```
# A tibble: 4 x 3
  path                      size format
  <chr>              <fs::bytes> <chr>
1 statcast_2020.rds        23.4M rds
2 statcast_2021.rds        63.1M rds
3 statcast_2022.rds          63M rds
4 statcast_2023.rds        63.8M rds
```

12.3.2 Using Apache Arrow and Apache Parquet

Apache Parquet is a file format that, when combined with the software framework Apache Arrow, provides a slick, scalable solution to the problem we raised above about how to store our data. The Parquet format is not as space-efficient as the *.rds format, but it is cross-platform and scalable, in that it will automatically chunk the data into partitions based on a grouping variable (in this case, the year). The **arrow** package for R provides a **dplyr** compatible interface that allows us to work with data in the Parquet format very easily. Because Arrow is columnar-oriented (as opposed to row-oriented), it can be very fast. Please see the chapter on Arrow in Wickham, Çetinkaya-Rundel, and Grolemund (2023) for more information about Arrow.

One nice feature of **arrow** is that it can read an entire directory of CSVs using
the `open_dataset()` function.

```
library(arrow)
sc_arrow <- statcast_dir |>
  open_dataset(format = "csv")
dim(sc_arrow)
```

```
[1] 2399921        92
```

Note that while the Arrow object `sc_arrow` behaves like a data frame—in
this case containing nearly 2.4 million rows!—it takes up almost no space in
memory.

```
sc_arrow |>
  object.size()
```

```
504 bytes
```

This is possible because the data is still on disk in the form of the CSVs. It
hasn't been read into R's memory yet. That won't stop us from querying the
data, however.

```
summary_arrow <- sc_arrow |>
  group_by(year = year(game_date), game_type) |>
  summarize(
    num_games = n_distinct(game_pk),
    num_pitches = n(),
    num_hr = sum(events == "home_run", na.rm = TRUE)
  )
summary_arrow |>
  collect()
```

```
# A tibble: 4 x 5
# Groups:   year [4]
  year game_type num_games num_pitches num_hr
  <int> <chr>        <int>       <int>  <int>
1 2020 R             898      263584   2304
2 2021 R            2429      709852   5944
3 2022 R            2430      708540   5215
4 2023 R            2430      717945   5868
```

So far we have just worked with an Arrow object that was backed by a directory
of CSVs. We can write the Arrow data frame in the Parquet format using the
`write_dataset()` function. Note that since we used the `group_by()` function
first, we'll get one Parquet file for each year. This is a form of file-based

partitioning that can provide significant performance advantages (Wickham, Çetinkaya-Rundel, and Grolemund 2023).

```
statcast_parquet <- path(data_dir, "statcast_parquet")
if (!dir.exists(statcast_parquet)) {
  dir.create(statcast_parquet)
}
sc_arrow |>
  group_by(year = year(game_date)) |>
  write_dataset(path = statcast_parquet, format = "parquet")
```

The `write_dataset()` function automatically creates a directory structure that partitions the data set into separate files of about 100 MB for each full season. This is just more than a quarter of the disk space occupied by the CSVs.

```
disk_space_parquet <- statcast_parquet |>
  dir_info(recurse = TRUE, glob = "*.parquet") |>
  select(path, size) |>
  mutate(
    format = "parquet",
    path = path_rel(path, start = statcast_parquet)
  )
disk_space_parquet
```

```
# A tibble: 4 x 3
  path                          size format
  <fs::path>              <fs::bytes> <chr>
1 year=2020/part-0.parquet      37.4M parquet
2 year=2021/part-0.parquet     101.5M parquet
3 year=2022/part-0.parquet     101.1M parquet
4 year=2023/part-0.parquet     102.4M parquet
```

12.3.3 Using DuckDB

While Arrow provides a **dplyr** interface that allows you to work seamlessly with Arrow objects in R as if they were data frames, Arrow is not SQL-based. Thus, while Arrow and Parquet are cross-platform, you would need another interface to write SQL queries against them.

Another fast, cross-platform alternative that *is* SQL-based is DuckDB. Like SQLite, DuckDB has a server-less architecture that can store data in memory or write its database files locally. This makes it a great option for someone who wants a SQL-interface but doesn't want to set up or maintain a SQL server. You can learn more about DuckDB in Wickham, Çetinkaya-Rundel, and Grolemund (2023).

DuckDB also implements a **dplyr** interface, so we set up a database connection in the same way we set up any **DBI**-compatible SQL database connection: using dbConnect(). However, in this case we want to write the database files to disk so that we can use them again in the future, and so that we can compare their sizes to the other storage formats. The dbdir argument specifies the path to a DuckDB database file that will be created for us if it doesn't exist already.

```
statcast_duckdb <- path(data_dir, "statcast_duckdb")
if (!dir.exists(statcast_duckdb)) {
  dir.create(statcast_duckdb)
}
library(duckdb)
con_duckdb <- dbConnect(
  drv = duckdb(),
  dbdir = path(statcast_duckdb, "statcast.ddb")
)
```

Initially, our DuckDB database doesn't contain any tables, so we use the dbWriteTable() function to copy the contents of the Arrow object to our DuckDB object.[1]

```
con_duckdb |>
  dbWriteTable("events", collect(sc_arrow), overwrite = TRUE)
```

Now, we can use our familiar **dplyr** interface to access the DuckDB database.

```
sc_ddb <- con_duckdb |>
  tbl("events")

summary_duckdb <- sc_ddb |>
  group_by(year = year(game_date), game_type) |>
  summarize(
    num_games = n_distinct(game_pk),
    num_pitches = n(),
    num_hr = sum(as.numeric(events == "home_run"), na.rm = TRUE)
  )

summary_duckdb
```

```
# Source:   SQL [4 x 5]
# Database: DuckDB v0.9.2 [bbaumer@Linux:R 4.3.2//statcast.ddb]
# Groups:   year
```

---

[1]The **arrow** package contains a function called to_duckdb() that will create a DuckDB object from an existing Arrow object, but we choose not to use that because in this case we want to copy the data.

	year	game_type	num_games	num_pitches	num_hr
	\<dbl\>	\<chr\>	\<dbl\>	\<dbl\>	\<dbl\>
1	2020	R	898	263584	2304
2	2022	R	2430	708540	5215
3	2021	R	2429	709852	5944
4	2023	R	2430	717945	5868

While both the **arrow** and **duckdb** packages provide **dplyr** interfaces, only **duckdb** works with SQL tools like dbGetQuery().

```
con_duckdb |>
  dbGetQuery("
    SELECT game_date, pitch_type, release_speed, pitcher
    FROM events
    WHERE release_speed > 100 AND events = 'home_run'
    LIMIT 6;
  ")
```

	game_date	pitch_type	release_speed	pitcher
1	2021-04-10	FF	100	594798
2	2021-05-18	SI	100	621237
3	2021-06-27	FF	100	543037
4	2021-06-29	SI	101	621237
5	2021-07-09	FC	100	661403
6	2021-07-16	FC	100	661403

The storage footprint of the DuckDB database is similar to that of the CSVs, but we will see in Section 12.4 that the performance is excellent.

```
disk_space_duckdb <- statcast_duckdb |>
  dir_info(recurse = TRUE, glob = "*.ddb") |>
  select(path, size) |>
  mutate(
    format = "duckdb",
    path = path_rel(path, start = statcast_duckdb)
  )
disk_space_duckdb
```

```
# A tibble: 1 x 3
  path                  size format
  <fs::path>    <fs::bytes> <chr>
1 statcast.ddb        1.28G duckdb
```

12.3.4   Using MySQL

Finally, we can use the MariaDB (MySQL) database that we set up in Section 11.2.

```
library(dbplyr)
library(RMariaDB)
con_mariadb <- dbConnect(MariaDB(), group = "abdwr")
```

Just as we did with DuckDB, we first copy the data to the MySQL server using dbWriteTable().

```
con_mariadb |>
  dbWriteTable("events", collect(sc_arrow), overwrite = TRUE)
```

Now we can query the database using the **dplyr** interface or write SQL queries.

```
sc_maria <- con_mariadb |>
  tbl("events")


summary_maria <- sc_maria |>
  group_by(year = year(game_date), game_type) |>
  summarize(
    num_games = n_distinct(game_pk),
    num_pitches = n(),
    num_hr = sum(events == "home_run", na.rm = TRUE)
  )
summary_maria
```

```
# Source:    SQL [4 x 5]
# Database: mysql  [abdwr@localhost:NA/abdwr]
# Groups:    year
  year game_type num_games num_pitches num_hr
  <int> <chr>        <int64>     <int64>  <dbl>
1 2020 R               898      263584   2304
2 2021 R              2429      709852   5944
3 2022 R              2430      708540   5215
4 2023 R              2430      717945   5868
```

Determining the storage footprint for a MySQL database is a bit more complicated, but the size of the events.ibd and events.frm files in the output below provide a lower bound on the size of the events table. The storage footprint here is about the same as those of the original CSV files and DuckDB.

```
disk_space_mariadb <- "/var/lib/mysql/abdwr/" |>
  dir_info(glob = "*events.*") |>
  select(path, size) |>
  mutate(
    format = "mariadb",
    path = path_rel(path, start = "/var/lib/mysql/abdwr/")
  )
disk_space_mariadb
```

```
# A tibble: 2 x 3
  path                   size format
  <fs::path> <fs::bytes> <chr>
1 events.frm       3.92K mariadb
2 events.ibd       1.22G mariadb
```

12.4 Performance Comparison

In this chapter, we have explored five different data storage formats (CSV, *.rds, Parquet, DuckDB, and MariaDB), as well as their corresponding R object interfaces, all of which are compatible with those of the **dplyr** package. Computing performance is often measured in terms of three quantities: computational speed, memory footprint, and disk storage footprint. We consider these three criteria in turn.

12.4.1 Computational speed

First, we compare the performance in terms of querying speed. We use the mark() function from the **bench** package to compare the amount of time it takes to compute the Statcast summary statistics for our 2020–2023 data. The five classes of objects are: 1) a tbl (data frame) which stores in the data in memory; 2) an Arrow object backed by the CSV files; 3) an Arrow object backed by the Parquet files, partitioned by year; 4) a DuckDB object; and 5) a MariaDB object.

First, we set up the tbl interface and query across all four years of data.

```
sc_tbl <- statcast_dir |>
  statcast_read_csv()

summary_tbl <- sc_tbl |>
  group_by(year = year(game_date), game_type) |>
  summarize(
```

```
    num_games = n_distinct(game_pk),
    num_pitches = n(),
    num_hr = sum(events == "home_run", na.rm = TRUE)
  )
```

Second, we set up an **arrow** object to read the from the Parquet files we created earlier, and to make use of the file-based partitioning based on year.

```
sc_arrow_part <- statcast_parquet |>
  open_dataset(partitioning = "year")

summary_arrow_part <- sc_arrow_part |>
  group_by(year = year(game_date), game_type) |>
  summarize(
    num_games = n_distinct(game_pk),
    num_pitches = n(),
    num_hr = sum(events == "home_run", na.rm = TRUE)
  )
```

Now we can benchmark the query execution times.

```
library(bench)
res <- mark(
  tbl = summary_tbl,
  arrow_csv = summary_arrow |> collect(),
  arrow_part = summary_arrow_part |> collect(),
  duckdb = summary_duckdb |> collect(),
  mariadb = summary_maria |> collect(),
  check = FALSE
) |>
  arrange(median)
res
```

```
# A tibble: 5 x 6
  expression       min    median   `itr/sec` mem_alloc `gc/sec`
  <bch:expr> <bch:tm>  <bch:tm>        <dbl> <bch:byt>    <dbl>
1 tbl          18.04ns      22ns 45777998.          0B        0
2 duckdb       59.18ms   59.38ms       16.6     446.1KB     5.52
3 arrow_part   143.9ms  145.94ms        6.79     47.6KB        0
4 arrow_csv   991.09ms  991.09ms        1.01     47.9KB        0
5 mariadb        3.54s     3.54s        0.282    154.8KB        0
```

The performance can vary greatly depending on the hardware available and software configuration of the computer. On this machine, which has 12 CPUs and 32 gigabytes of RAM, the results indicate that **duckdb** is significantly faster than the other databases. The `tbl` interface already has the data in memory, so of course it is by far the fastest. In this case, our query ran across all four partitions, so the partitioning scheme of Parquet wasn't that useful. Still, the Arrow object that was backed by Parquet files was about 7 times faster than the Arrow object backed by the CSV files. However, the DuckDB instance was still about 2 times faster than the Arrow/Parquet object. Even the Arrow object backed by the CSV files was about 4 times faster than **RMariaDB**, which was by far the worst performer.

In Section 12.5, we explore whether the performance of the Arrow/Parquet object will improve when querying the data for individual seasons separately.

12.4.2 Memory footprint

Second, we note that while the other objects occupy a negligible amount of space in memory, the `tbl` object takes up 1.6 Gb of RAM! This is because, as noted above, the `tbl` object stores the data in memory, while the other objects leave the data on disk and only query the relevant data when prompted. The size of these objects *in bytes* is shown below. Thus, the superior performance of the `tbl` object comes at a cost.

```
list(
  "tbl" = sc_tbl,
  "arrow" = sc_arrow,
  "duckdb" = sc_ddb,
  "mariadb" = sc_maria
) |>
  map_int(object.size)
```

tbl	arrow	duckdb	mariadb
1752447264	504	44384	12672

12.4.3 Disk storage footprint

Third, in terms of data storage space, the `*.rds` format is the most compact. However, it comes with its own limitations: `*.rds` only works with R. The Parquet files take up about 50% more space than the `*.rds` files, but they work across platforms and they implement seamless partitioning. The DuckDB files take up about three times as much disk space as the Parquet files (and this is before we have built any indexes!), but they are faster to query.

```
disk_space <- bind_rows(
  disk_space_csv,
  disk_space_rds,
  disk_space_parquet,
  disk_space_duckdb,
```

```
    disk_space_mariadb
  )
  disk_space |>
    group_by(format) |>
    summarize(footprint = sum(size)) |>
    arrange(desc(footprint))
```

```
# A tibble: 5 x 2
  format       footprint
  <chr>      <fs::bytes>
1 duckdb          1.28G
2 mariadb         1.22G
3 csv             1.22G
4 parquet       342.46M
5 rds           213.36M
```

12.4.4 Overall guidelines

In our experiment with a 2.4 million row data set, on this particular computer using these particular data, we confirmed the obvious: reading the data into a `tbl` object in memory leads to the fastest computational performance while requiring the largest footprint in memory. Among the interfaces that read the data from disk, **duckdb** provided the fastest computational performance, while **RMariaDB** offered the slowest performance. Neither reduced the largest footprint on disk substantially from the original CSV files. Arrow using the Parquet storage format provided medium performance with medium footprint on disk. R's internal `*.rds` storage format was the most compact, but the least versatile.

This leads us to the following guidelines as we consider these options in broader practice.

- If your data is small (i.e., less than a couple hundred megabytes), just use CSV because it's easy, cross-platform, and versatile. The fact that it is not space-efficient won't matter because your data is small anyway.
- If your data is larger than a couple hundred megabytes and you're just working in R (either by yourself or with a few colleagues), use `.rds` because it's space-efficient and optimized for R. (Note that this is how we chose to store much of the Retrosheet and Statcast data we use in this book.) Reading these files into `tbl` objects will lead to fast performance, and the data probably aren't large enough to eat up enough of your computer's memory for you to notice.
- If your data is around a gigabyte or more and you need to share your data files across different platforms (i.e., not just R but also Python, etc.) and you don't want to use a SQL-based RDBMS, store your data in the Parquet format and use the **arrow** package. Parquet is cross-platform and Arrow

scales better than .rds. Both the performance and the storage footprint are likely to meet your needs. The file-based partitioning scheme may or may not help you, depending on how you are querying the database.

- If you want to work in SQL with a local data store, use DuckDB, because it offers more features and better performance than RSQLite, and doesn't require a server-client architecture that can be cumbersome to set up and maintain.
- If you have access to a RDBMS server (hopefully maintained by a professional database administrator), use the appropriate **DBI** interface (e.g., **RMariaDB**, **RPostgreSQL**, etc.) to connect to it. A well-oiled server with sufficient resources will easily outperform anything you can do on your personal computer.

Making the appropriate choice for your situation will depend on weighing these factors carefully.

12.5 Launch Angles and Exit Velocities, Revisited

In the previous edition of this book, we created a data graphic that showed how wOBA varied as a function of launch angle and exit velocity for all batted balls in the 2017 season. That data graphic appeared on the cover of the second edition. Here, we produce a similar graphic across our four years of data. In doing so, we revisit the performance of our fastest data interfaces, and compare how well they work in practice.

Recall that while DuckDB stores its data in one large file, Arrow employs a file-based partitioning scheme that writes our data to a separate file for each year. Consider what happens when we ask the database to give us all the data for a particular player, say Pete Alonso, in a particular year, say 2020. DuckDB has to look in the whole big file for these data, but Arrow only has to look at the file for 2020, which in this case is less than one quarter the size of all of the files together. Then it only needs to look for Alonso's data within that file. Because the file is smaller, it should be faster to find the relevant data.

Conversely, if we were to ask for Alonso's data across all four years, the file-based partitioning wouldn't do any good, because we'd have to consult all of the files anyway.

The following function pulls the data we need for any given batter and set of years.

```
read_bip_data <- function(tbl, begin, end = begin,
                          batter_id = 624413) {
```

```
x <- tbl |>
  mutate(year = year(game_date)) |>
  group_by(year) |>
  filter(type == "X", year >= begin, year <= end) |>
  select(
    year, game_date, batter, launch_speed, launch_angle,
    estimated_ba_using_speedangle,
    estimated_woba_using_speedangle
  )
if (!is.null(batter_id)) {
  x <- x |>
    filter(batter == batter_id)
}
x |>
  collect()
}
```

First, we compare the computational performance for pulling Pete Alonso's balls in play for a single season, 2020.

```
mark(
  tbl = nrow(read_bip_data(sc_tbl, 2020)),
  arrow = nrow(read_bip_data(sc_arrow_part, 2020)),
  duckdb = nrow(read_bip_data(sc_ddb, 2020)),
  iterations = 5
) |>
  arrange(median)
```

```
# A tibble: 3 x 6
  expression        min    median `itr/sec` mem_alloc `gc/sec`
  <bch:expr> <bch:tm>  <bch:tm>      <dbl> <bch:byt>     <dbl>
1 duckdb        91.8ms    95.6ms       9.70    1.07MB      3.88
2 arrow        218.3ms     221ms       4.44  408.14KB      5.33
3 tbl          390.6ms   394.9ms       2.49  280.27MB     0.498
```

Amazingly, both the **duckdb** and **arrow** objects outperform the `tbl` object, which is remarkable given that the `tbl` object is stored in R's memory and the others are reading from files stored on disk. This result reveals how well these highly optimized technologies work in practice. It's also worth noting that while the **duckdb** object is still about twice as fast as the **arrow** object, the latter greatly improved its performance relative to the previous comparison. This is because the file-based partitioning scheme was useful in this case, because we were only querying for data in an individual year.

If instead, we query across years, this performance boost disappears, and **duckdb** once again several times faster than **arrow**.

```
mark(
  tbl = nrow(read_bip_data(sc_tbl, 2021, 2023)),
  arrow = nrow(read_bip_data(sc_arrow_part, 2021, 2023)),
  duckdb = nrow(read_bip_data(sc_ddb, 2021, 2023)),
  iterations = 5
) |>
  arrange(median)
```

```
# A tibble: 3 x 6
  expression        min   median `itr/sec` mem_alloc `gc/sec`
  <bch:expr> <bch:tm> <bch:tm>      <dbl> <bch:byt>     <dbl>
1 duckdb         89.7ms    100ms       9.35     983KB      3.74
2 arrow         398.7ms    407ms       2.39     375KB      2.87
3 tbl           548.1ms    552ms       1.76     520MB      0.352
```

12.5.1 Launch angles over time

We're now ready to make our plot, and compare the relationship of wOBA to both launch angle and exit velocity, but now over time. Just as we did in the second edition, we add some helpful guidelines to our plot.

```
guidelines <- tibble(
  launch_angle = c(10, 25, 50),
  launch_speed = 40,
  label = c("Ground balls", "Line drives", "Flyballs")
)
```

Since **duckdb** proved to be our best performer, we'll use it to pull the data and draw the plot. Note that we use the `slice_sample()` function to avoid plotting **all** the data.

```
ev_plot <- sc_ddb |>
  read_bip_data(2020, 2023, batter_id = NULL) |>
  # for speed
  slice_sample(prop = 0.2) |>
  ggplot(
    aes(
      x = launch_speed,
      y = launch_angle,
      color = estimated_woba_using_speedangle
    )
  ) +
  geom_hline(
    data = guidelines, aes(yintercept = launch_angle),
    color = "black", linetype = 2
```

```
) +
geom_text(
  data = guidelines,
  aes(label = label, y = launch_angle - 4),
  color = "black", hjust = "left"
) +
geom_point(alpha = 0.05) +
scale_color_viridis_c("BA") +
scale_x_continuous(
  "Exit velocity (mph)",
  limits = c(40, 120)
) +
scale_y_continuous(
  "Launch angle (degrees)",
  breaks = seq(-75, 75, 25)
) +
facet_wrap(vars(year))
```

A few observations stand out in Figure 12.1. First, line drives have a very high probability of being a hit, but the likelihood depends on how hard the ball comes off the bat and how high it is going. Second, there is a "sweet spot" where nearly all batted ball are hits. These form a white pocket centered around a launch angle of about 25 degrees and an exit velocity of at least 100 mph. As we will see later, these are often home runs. The contention is that batters are optimizing their swings to produce batted balls with these properties.

```
ev_plot +
  guides(color = guide_colorbar(title = "wOBA"))
```

Do you see any differences across the various years? Other than the fact that there is less data in 2020 due to the pandemic-shortened season, the relationship appears to be about the same.

12.6 Further Reading

The **abdwr3edata** package contains all of the functions displayed in this chapter. Chapter 11 of this book covers the use of a MySQL database to explore park factors. In Section 11.2, we illustrate how to setup and use a MySQL server, and in Section 11.6, we discuss how to build your own baseball database.

Chapter 21 of Wickham, Çetinkaya-Rundel, and Grolemund (2023) covers databases, and explicates the **dplyr** interface to **duckdb** and other

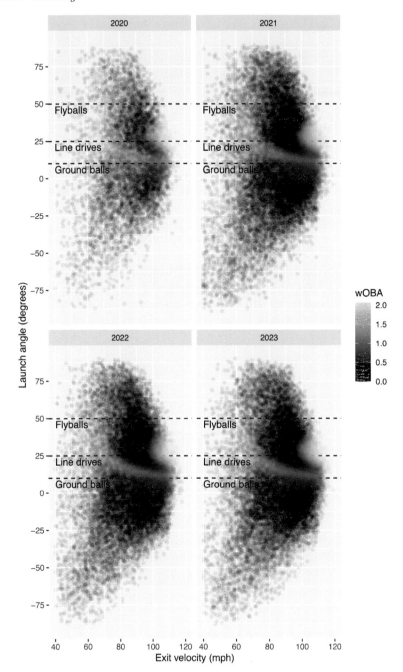

FIGURE 12.1
Estimated wOBA as a function of launch angle and exit velocity, 2020–2023.

DBI-compliant databases. Chapter 22 discusses Arrow and Parquet in greater detail.

12.7   Exercises

1. Home runs on fast pitches

Use the database constructed in this chapter to find the total number of home runs that were hit on fastballs traveling in excess of 100 mph across the four seasons 2020–2023. What percentage of the total number of home runs hit in each season were hit on such fastballs?

2. Stolen base percentages by velocity

Use the database constructed in this chapter to compute the stolen base success rate for every pitch speed, rounded to the nearest 1 mph. Separate your analysis by steals of 2nd base versus 3rd base. Does pitch speed appear to be correlated with stolen base success rate?

13

Home Run Hitting

13.1 Introduction

Home run hitting has always been associated with the great sluggers in baseball history, such as Babe Ruth, Roger Maris, Hank Aaron and Barry Bonds. But since the introduction of Statcast data in 2015, there has been an explosion in home run hitting with a remarkable 6776 home runs hit in the 2019 season. That raises the general question of what variables contribute to home run hitting. Many of the possible explanatory variables for the increase in home run hitting are contained in the Statcast data.

This chapter focus on the exploration of home runs using Statcast data on balls in play from the 2021 and 2023 seasons. Section 13.2 describes the variables from a subset of the Statcast data. Launch variables such as launch angle, exit velocity and spray angle are collected for each ball put in play and this section explores the association of these variables with home runs using data from the 2023 season. Team abbreviations for the home and away teams can be used to compute ballpark effects. The identities of the pitcher and batter are collected for each ball in play and by use of random effects model, one can see if the variation in home run rates is attributed to the batters or to the pitchers.

There is much interest in the potential causes behind changes home run hitting across seasons (see Albert et al. (2018), Albert et al. (2019)). Section 13.3 focuses on the changes between the 2021 and 2023 seasons. Although on the surface there appears to be little difference in home runs hit, this section shows how one can detect changes, both in batter behavior and in the carry properties of the baseball.

DOI: 10.1201/9781032668239-13

13.2   Exploring a Season of Home Run Hitting

13.2.1   Meet the data

We begin by reading in the data file `sc_bip_2021_2023.rds` which contains information on all balls put in play for the 2021 and 2023 seasons. Please see Section 12.2 for details on how to acquire these data. Using the `year()` function in the **lubridate** package, we define a `Season` variable, and we define `HR` to be 1 if a home run is hit, and 0 otherwise. The `filter()` function is used to define `sc_2023`, the balls in play for the 2023 season. The `glimpse()` function is used to identify the data type for all 14 variables in this Statcast dataset.

```
sc_two_seasons <- here::here("data/sc_bip_2021_2023.rds") |>
  read_rds() |>
  mutate(
    Season = year(game_date),
    HR = ifelse(events == "home_run", 1, 0)
  )
sc_2023 <- sc_two_seasons |>
  filter(Season == 2023)
glimpse(sc_2023)
```

```
Rows: 124,199
Columns: 16
$ game_pk         <dbl> 718773, 718774, 718778, 718773, 718781~
$ game_date       <date> 2023-03-30, 2023-03-30, 2023-03-30, 2~
$ batter          <dbl> 613564, 643446, 453568, 641584, 527038~
$ pitcher         <dbl> 656605, 645261, 605483, 656605, 543037~
$ events          <chr> "triple", "single", "single", "single"~
$ stand           <chr> "L", "L", "L", "L", "R", "R", "R", "R"~
$ p_throws        <chr> "R", "R", "L", "R", "R", "L", "R", "R"~
$ hit_distance_sc <dbl> 134, 9, 162, 254, 51, 56, 42, 185, 171~
$ hc_x            <dbl> 215.1, 164.8, 90.7, 196.9, 110.2, 153.~
$ hc_y            <dbl> 107.2, 105.1, 133.9, 95.2, 148.4, 209.~
$ launch_speed    <dbl> 94.2, 93.7, 59.1, 111.7, 94.8, 69.5, 1~
$ launch_angle    <dbl> 9, -19, 27, 13, 1, 81, -2, 9, 9, 7, 76~
$ home_team       <chr> "CIN", "MIA", "SD", "CIN", "NYY", "SD"~
$ away_team       <chr> "PIT", "NYM", "COL", "PIT", "SF", "COL~
$ Season          <dbl> 2023, 2023, 2023, 2023, 2023, 2023, 20~
$ HR              <dbl> 0, 0, 0, 0, 0, 0, 0, 0, 0, 0, 0, 0, 0,~
```

The `game_date` variable provides the date of the game and the `batter` and `pitcher` are numerical codes for the identity of the batter and pitcher respectively for the particular ball put into play. The `events` variable is a description of the outcome of the ball in play. The `stand` and `p_throws` variables give the sides (L or R) for the batter and pitcher. The `hit_distance_sc` variable is the distance (in feet) that the ball goes before it hits the ground. The `hc_x` and `hc_y` variables gives the location of ball put in the field. The `launch_speed` and `launch_angle` variables give the exit velocity (in mph) and launch angle (in degrees) of the ball when it leaves the bat. The `home_team` and `away_team` variables give abbreviations of the home and visiting teams in the game. See Appendix C for more details on the variables available through Statcast.

13.2.2 Home runs and launch variables

To begin, we use the `summarize()` and `mutate()` functions to compute the number of balls in play, the home run count and the home run rate in the 2023 season.

```
S <- sc_2023 |>
  summarize(
    BIP = n(),
    HR = sum(events == "home_run")
  ) |>
  mutate(HR_Rate = 100 * HR / BIP)
S
```

```
# A tibble: 1 x 3
     BIP    HR HR_Rate
   <int> <int>   <dbl>
1 124199  5868    4.72
```

We see that 4.72 percent of batted balls were home runs.

We use a model to understand the relationship between home run hitting and the launch angle and launch speed variables. The generalized additive model of the form

$$\log \left(\frac{\Pr(HR)}{1 - \Pr(HR)} \right) = s(LA, LS)$$

is fit where $\Pr(HR)$ is the probability of a home run, LA and LS denote the launch angle and launch speed measurements, and $s()$ is a smooth function of the two measurements.

Recall the variable `HR` is defined to be 1 if a home run is hit, and 0 otherwise. We fit the generalized additive model by use of the `gam()` function from the **mgcv** package.

```
library(mgcv)
fit_23 <- gam(
  HR ~ s(launch_angle, launch_speed),
  family = binomial,
  data = sc_2023
)
```

This model can be used to predict the home run probability for any values of the launch variables. For example, suppose a batter hits a ball at 105 mph with a launch angle of 25 degrees. We use the `predict()` function with the `type` argument set to `response` to predict the home run probability. We see from the output that a batted ball hit at 105 mph with a launch angle of 25 degrees has a 77.2% chance of being a home run.

```
fit_23 |>
  predict(
    newdata = data.frame(
      launch_speed = 105,
      launch_angle = 25
    ),
    type = "response"
  )
```

```
      1
0.772
```

One can display these home run predictions by use of a filled contour graph. Using the `expand_grid()` function, we define a 50 by 50 grid of launch variables where the launch angle is between 15 and 40 angles and the launch speed is between 90 and 100 mph. We use the `predict()` function to compute home run probability predictions on this grid.

```
grid <- expand_grid(
  launch_angle = seq(15, 40, length = 50),
  launch_speed = seq(90, 110, length = 50)
)
hats <- fit_23 |>
  predict(newdata = grid, type = "response")
grid <- grid |>
  mutate(prob = hats)
```

Using the `geom_contour_fill()` function from the **metR** package, we construct a contour graph displayed in Figure 13.1, where the contour lines are at the values from 0.1 to 0.9 in steps of 0.2.

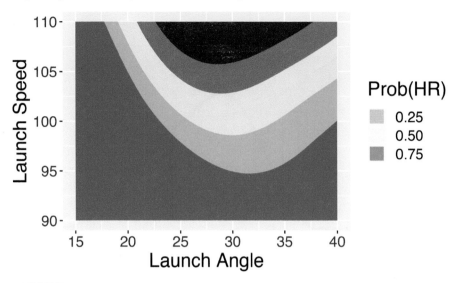

FIGURE 13.1
Filled contour graph of home run probability as a function of the launch angle and launch velocity from 2023 data.

```
library(metR)
ggplot(grid) +
  geom_contour_fill(
    aes(
      x = launch_angle,
      y = launch_speed,
      z = prob
    ),
    breaks = c(0, .1, .3, .5, .7, .9, 1),
    linewidth = 1.5
  ) +
  scale_fill_distiller(palette = "Spectral") +
  theme(text = element_text(size = 18)) +
  labs(x = "Launch Angle", y = "Launch Speed") +
  guides(fill = guide_legend(title = "Prob(HR)"))
```

Certainly, high launch speeds are a contributing factor in home run hitting. But certainly the launch angle is also an important variable. From the contour graph, we see that the chance of home run at 105 mph and 20 degrees is approximately the same as the probability when the launch speed is 95 and the launch angle is 30. Batted balls hit over 105 mph with a launch angle between 25 and 35 degrees are likely to be home runs.

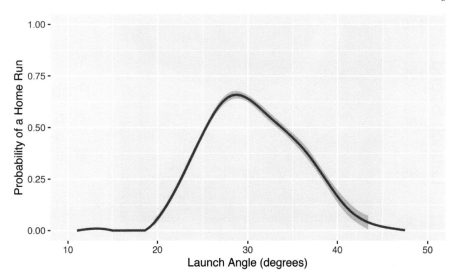

FIGURE 13.2
Smoothed home run probability as a function of launch angle for batted balls
hit between 100 and 105 mph.

13.2.3 What is the optimal launch angle?

The previous graph raises the question—what is the optimal launch angle for
hitting a home run? To address this question, let's focus on batted balls hit
between 100 and 105 mph. Using the `geom_smooth()` function from the **ggplot2**
package we display the smoothed home run probability in Figure 13.2 when
HR is graphed against `launch_angle` for these batted balls. The message from
this graph is, for these hard-hit balls, the home run probability is maximized
at 28 degrees.

```
sc_2023 |>
  filter(launch_speed >= 100, launch_speed <= 105) |>
  ggplot(aes(x = launch_angle, y = HR)) +
  geom_smooth(method = "gam") +
  scale_y_continuous(
    "Probability of a Home Run",
    limits = c(0, 1)
  ) +
  scale_x_continuous(
    "Launch Angle (degrees)",
    limits = c(10, 50)
  )
```

13.2.4 Temperature effects

Another contributing factor to home run hitting is temperature. Generally the ball carries further in warmer temperatures. To see this effect, we need to collect some additional data. Using the `mlb_game_info()` function from the **baseballr** package, we collect the game-time temperature and park name for all games in the 2023 season—these data are stored in the data frame `temps_2023` in the **abdwr3edata** package. Since the temperature is only variable for parks that don't have a dome, we collect another data frame `parks_2023` that contains the park name and the dome status for all parks used in the 2023 season.

We read in these two additional data files and by two applications of the `inner_join()` function, merge them with the Statcast data.

```
library(abdwr3edata)
temps_parks_2023 <- temps_2023 |>
  inner_join(parks_2023, by = c("Park"))
sc_2023 <- sc_2023 |>
  inner_join(temps_parks_2023, by = "game_pk")
```

We use the `filter()` function to restrict attention to games where the `Dome` variable is "No". Then we use the `group_by()` and `summarize()` functions to compute the count of balls in play and home runs for each day in the 2013 season.

```
temp_hr <- filter(sc_2023, Dome == "No") |>
  group_by(temperature) |>
  summarize(
    BIP = n(),
    HR = sum(HR, na.rm = TRUE)
  ) |>
  mutate(HR_Rate = 100 * HR / BIP)
```

Using **ggplot2**, we construct a scatterplot of the home run rate against the temperature and overlay a smoothing line in Figure 13.3. We see that there is a general tendency for the home run rate to rise for increasing game-time temperature.

```
temp_hr |>
  filter(temperature >= 55, temperature <= 90) |>
  ggplot(aes(temperature, HR_Rate)) +
  geom_point() +
  geom_smooth(method = "lm", formula = "y ~ x") +
  labs(
    x = "Temperature (deg F)",
    y = "Home Run Rate"
  )
```

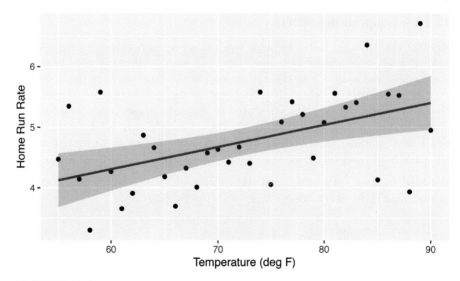

FIGURE 13.3
Scatterplot of home run rate against temperature during the 2023 season.

To measure this temperature effect, we fit a least-squares model of the form

$$HR\,Rate = \beta_0 + \beta_1(Temp - 70)$$

```
lm(HR_Rate ~ I(temperature - 70), weights = BIP, data = temp_hr) |>
  pluck(coef)
```

```
(Intercept) I(temperature - 70)
   4.652                 0.041
```

From the output, we predict the home run rate to be 4.65% at a game-time temperature of 70 degrees and the rate increases by 0.041 for each additional one-degree Fahrenheit increase in temperature.

13.2.5   Spray angle effects

Besides the launch speed and launch angle, we have measurements that can be used to construct the spray angle, which is the radial angle that is set to 0 for balls hit up the middle, −45 degrees for balls hit along the third base line and 45 degrees for balls hit along the first base line. Statcast doesn't provide the spray angle measurement directly, but one can compute this angle from the location of the batted ball measurements `hc_x` and `hc_y`. Please see Section C.7 for a detailed explanation of the calculation. Using the `mutate()` function, we compute the `spray_angle`, measured in degrees and define `sc_hr` to be the Statcast measurements for the home runs hit in the 2023 season.

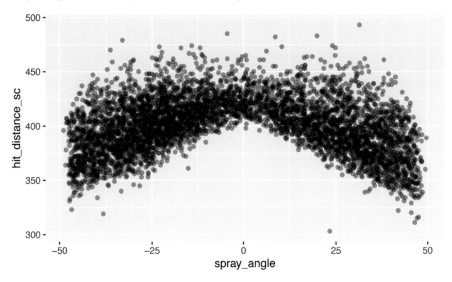

FIGURE 13.4
Scatterplot of spray angle and distance traveled for home runs hit in the 2023
season.

```
sc_2023 <- sc_2023 |>
  mutate(
    location_x = 2.5 * (hc_x - 125.42),
    location_y = 2.5 * (198.27 - hc_y),
    spray_angle = atan(location_x / location_y) * 180 / pi
  )
sc_hr <- sc_2023 |>
  filter(events == "home_run")
```

We construct a scatterplot of `spray_angle` and batted ball distance variable
`hit_distance_sc` shown in Figure 13.4. Since the distances to the fences are
greatest in center field, the batted ball distance tends to be greatest for home
runs hit for spray angle values near zero and the distances tend to be smallest
for balls hit along the first and third base lines. From the graph, we identify
five home runs that had a distance exceeding 475 feet. There is one curious
outlier—a home run with a distance of 300 feet and spray angle close to 25
degrees. This turns out to be an inside-the-park home run hit by Bobby Witt,
Jr. of the Kansas City Royals.

```
ggplot(sc_hr, aes(spray_angle, hit_distance_sc)) +
  geom_point(alpha = 0.25)
```

Batters tend to hit home runs in the "pull" direction. We can verify this
graphically. In Figure 13.5, we divide the hitters by the batting side, either

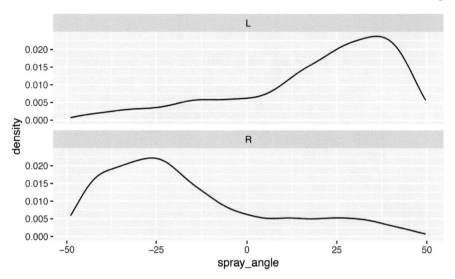

FIGURE 13.5
Density estimates of spray angle of home runs hit by left- and right-handed hitters.

L or H, and then display density estimates of the spray angle for each group of hitters. As expected, left-handed hitters tend to hit home runs to the right (positive spray angle) and right-handed hitters tend to hit home runs in the negative spray angle direction. What is interesting is that the degree of pull is strongest for the left-handed batters—the modal (most likely) spray angle value is 37 degrees for left-handed hitters compared to –25 degrees for right-handed hitters.

```
ggplot(sc_hr, aes(spray_angle)) +
  geom_density() +
  facet_wrap(vars(stand), ncol = 1)
```

13.2.6 Ballpark effects

As baseball fans know, all ballparks are not the same with respect to home run hitting. The parks differ with respect to the distances to the fences and climate, so it is easier to hit home runs in some MLB parks. One way to define a ballpark effect with respect to home runs for, say, Atlanta, is to compare the ratio of all home runs hit by the Braves and their opponents at Truist Park and all home runs hit by the Braves and opponents when the Braves are on the road. (See Section 11.7 for in-depth example of how to calculate ballpark factors.)

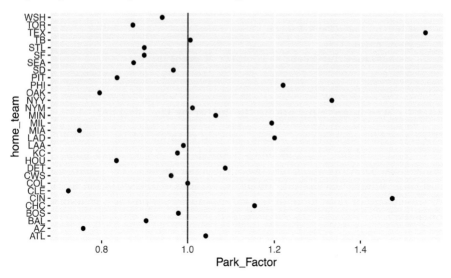

FIGURE 13.6
Dotplots of home run ballpark effects from 2023 season data.

One can compute home run ball park effects for all 30 teams by use of the `home_team` and `away_team` variables. Using the `group_by()` and `summarize()` functions, we compute the count of home runs hit for all teams (and opponents) when they are at home. Similarly, we compute the home run count for all teams when they are on the road. The `Home` and `Away` data frames are merged using the `inner_join()` function and the variable `Park_Factor` is used to define the park factor.

```
sc_home <- sc_2023 |>
  group_by(home_team) |>
  summarize(HR = sum(HR))
sc_away <- sc_2023 |>
  group_by(away_team) |>
  summarize(HR = sum(HR))
pf <- sc_home |>
  inner_join(sc_away, join_by(home_team == away_team)) |>
  mutate(Park_Factor = HR.x / HR.y)
```

These park factor values are graphed using dotplots in Figure 13.6, where a vertical line at 1 is added that shows a "neutral" ballpark. Several parks stand out as being home run friendly, the Texas Rangers (TEX) and New York Yankees (NYY) have park factors exceeding 1.5, while the Cincinnati Reds (CIN) have a park factor close to 1.4. It should be said that park factors for a single season tend to be unstable and so Baseball Savant will typically display park factors over a three-season period.

```
ggplot(pf, aes(Park_Factor, home_team)) +
  geom_point() +
  geom_vline(xintercept = 1, color = crcblue)
```

13.2.7 Are home runs about the hitter or the pitcher?

When we think of home run hitting, we tend to think of batters with high home run counts and not think as much about pitchers who allow many (or few) home runs. That raises the question—when one looks at the variation of home run rates among batters and pitchers, is more of the variation attributable to the hitters or does more of the variation come from the variability among pitchers?

This question can be addressed by use of a non-nested random effects model. Let p_{ij} denote the probability that a batted ball is a home run depending on the ith batter and the jth pitcher. We consider the logistic random effects model written as

$$\log\left(\frac{p}{1-p}\right) = \mu + \beta_i + \gamma_j$$

where μ is an overall effect, β_i is the effect due to the ith hitter and γ_j is the effect due to the jth pitcher. We let the batter effects follow a normal distribution with mean 0 and standard deviation σ_b and the pitcher effects be normal with mean 0 and standard deviation σ_p.

This random effects model is conveniently fit using the `glmer()` function from the **lme4** package. The syntax for the formula is similar to `lm()` function where the | indicates that the pitcher and the batter identities are random effects. We are interested in the estimated values of the random effects standard deviations σ_b and σ_p that are displayed using the `VarCorr()` function.

```
library(lme4)
fit <- glmer(
  HR ~ (1 | pitcher) + (1 | batter),
  data = sc_2023,
  family = binomial
)
VarCorr(fit)
```

```
Groups  Name        Std.Dev.
pitcher (Intercept) 0.188
batter  (Intercept) 0.523
```

Note that the estimated value of σ_b is much larger than the estimated σ_p indicating that most of the variation in home run rates is attributable to the hitter, not the pitcher. So our instincts are correct—it is best to focus on leaderboards on home run hitting instead of home runs allowed by pitchers.

13.3 Comparing Home Run Hitting in the 2021 and 2023 Seasons

13.3.1 Introduction

We are interested in comparing home run hitting for the 2021 and 2023 seasons. On the surface, home run hitting for the two seasons was similar since 1.21 home runs were hit per game per game in 2023 compared with 1.22 in 2021. But we will see that there have been changes both in batter behavior and in the carry properties of the baseball between the two seasons, and we'll explore these changes by comparing specific rates.

From Figure 13.1, we see that most home runs are hit when the launch angle is between 20 and 40 degrees and the exit velocity is between 95 and 110 mph. In this section, we divide this launch variable space into subregions and consider two rates computed on each subregion. We consider the BIP rate

$$BIP\,Rate = 100 \times \frac{BIP}{N},$$

the percentage of batted balls BIP hit in the subregion where N is the total number of batted balls. Also we consider the percentage of home runs hit in the subregion

$$HR\,Rate = 100 \times \frac{HR}{BIP}.$$

Section 13.3.2 describes the process of binning the launch variable space into subregions and Section 13.3.3 describes how one can plot measures over the bins. Sections 13.3.4 and 13.3.5 explain how one compares rates across two seasons using a logit reexpression. Associated graphs show how players are changing their batting behavior and how the ball's carry properties have changed from 2021 to 2023.

13.3.2 Binning launch variables

We write a special function `bin_rates()` to compute the balls-in-play and home run rates across regions of the launch variable space. One inputs the Statcast data frame `sc_ip`, a vector of breakpoints for launch angle `LA_breaks` and a vector of breakpoints for the exit velocity `LS_breaks`.

The `mutate()` and `cut()` functions are used to categorize the launch speeds and launch angles using the breakpoints vectors. Then by use of the `group_by()` function, the number of balls in play and the home run count is computed for each bin. The `mutate()` function is used again to compute the BIP and home run rates for all bins.

```
bin_rates <- function(sc_ip, LA_breaks, LS_breaks) {
  Total_BIP <- nrow(sc_ip)
  sc_ip |>
    mutate(
      LS = cut(launch_speed, breaks = LS_breaks),
      LA = cut(launch_angle, breaks = LA_breaks)
    ) |>
    filter(!is.na(LA), !is.na(LS)) |>
    group_by(LA, LS) |>
    summarize(
      BIP = n(),
      HR = sum(HR),
      .groups = "drop"
    ) |>
    mutate(
      BIP_Rate = 100 * BIP / Total_BIP,
      HR_Rate = 100 * HR / BIP
    )
}
```

To illustrate the use of `bin_rates()`, we use values of launch angle equally spaced from 20 to 40 degrees and values of launch velocity spaced from 95 to 110 mph. The output of `bin_rates()` is a data frame with the launch angle and launch speed intervals `LA`, `LS`, the counts `BIP`, `HR`, and the rates `BIP_Rate` and `HR_Rate`. We display the first six rows of this data frame.

```
LA_breaks <- seq(20, 40, by = 5)
LS_breaks <- seq(95, 110, by = 5)
S <- sc_2023 |>
  bin_rates(LA_breaks, LS_breaks)
slice_head(S, n = 6)
```

```
# A tibble: 6 x 6
  LA        LS           BIP    HR BIP_Rate HR_Rate
  <fct>     <fct>      <int> <dbl>    <dbl>   <dbl>
1 (20,25]   (95,100]    1459    64     1.30    4.39
2 (20,25]   (100,105]   1572   459     1.40    29.2
3 (20,25]   (105,110]    932   646    0.828    69.3
4 (25,30]   (95,100]    1494   311     1.33    20.8
5 (25,30]   (100,105]   1382   861     1.23    62.3
6 (25,30]   (105,110]    684   637    0.608    93.1
```

13.3.3 Plotting measure over bins

We write the function `bin_plot()` to display a measure over bins over the launch variable space. The inputs are the data frame `S` containing the bin frequencies and rates, the vector of breakpoints LA_breaks, LS_breaks, and the measure to be displayed label.

As a first step, we write a function compute_bin_midpoint() to compute the midpoints of the intervals. The parse_number() function from the **readr** package strips away unwanted characters.

```
compute_bin_midpoint <- function(x) {
  x |>
    as.character() |>
    str_split_1(",") |>
    map_dbl(parse_number) |>
    mean()
}
```

We use the function geom_text() to display label over the bins and applications of geom_vline() and geom_hline() are used to overlay lines at the bin boundaries.

```
bin_plot <- function(S, LA_breaks, LS_breaks, label) {
  S |>
    mutate(
      la = map_dbl(LA, compute_bin_midpoint),
      ls = map_dbl(LS, compute_bin_midpoint)
    ) |>
    ggplot(aes(x = la, y = ls)) +
    geom_text(aes(label = {{label}}), size = 8) +
    geom_vline(
      xintercept = LA_breaks,
      color = crcblue
    ) +
    geom_hline(
      yintercept = LS_breaks,
      color = crcblue
    ) +
    theme(text = element_text(size = 18)) +
    labs(x = "Launch Angle", y = "Launch Speed")
}
```

We illustrate the use of bin_plot() on the 2023 Statcast data in Figure 13.8. We use the data frame S containing the binned frequencies and indicate that HR is the variable to display. This figure shows that 861 home runs were hit where the launch angle is in (25, 30) and the exit velocity is in (100, 105).

```
bin_plot(S, LA_breaks, LS_breaks, HR)
```

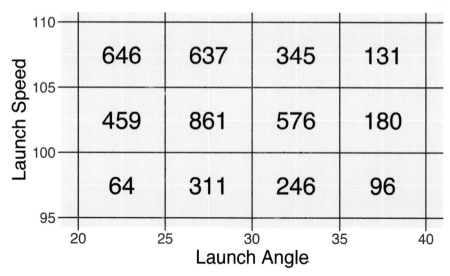

FIGURE 13.7
Home run counts over different regions defined by launch angle and launch
speed for 2023 data.

Instead suppose we wish to display the home run rate `HR_Rate`. In Figure 13.8,
we see the chance that a batted ball with launch angle between 25 and 30
degrees and exit velocity between 100 and 105 mph has a 62% chance of being
a home run. As one would expect, the home run rates increase for larger values
of launch speed.

```
S |>
  bin_plot(
    LA_breaks, LS_breaks,
    label = paste(round(HR_Rate, 0), "%", sep = "")
  )
```

13.3.4   Changes in batter behavior

We apply the functions `bin_rates()` and `bin_plot()` to compare batted ball
and home run rates between the 2021 and 2023 seasons. Using the same vectors
of breakpoints for launch angle and launch speed, we bin the 2021 and 2023
Statcast data frames using two applications of `bin_rates()` via `map()`. Then
we combine these two summary data frames using the `list_rbind()` function.

```
S2 <- sc_two_seasons |>
  group_split(Season) |>
```

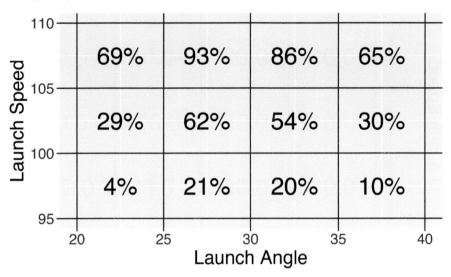

FIGURE 13.8
Home run rates over different regions defined by launch angle and launch speed for 2023 data.

```
map(bin_rates, LA_breaks, LS_breaks) |>
set_names(c(2021, 2023)) |>
list_rbind(names_to = "year")
```

We want to compare the balls in play rates for 2021 and 2023. One issue with rates is that rates near 0 and 100 percent tend to have smaller variation than rates near 50 percent. One way of addressing this variation issue is through the use of logits. If R denotes a rate on the percentage scale, then we define the logit of R to be:

$$logit(R) = \log\left(\frac{R}{100 - R}\right)$$

Logits tend to have similar spread across all rate values. If we have two rates measured for two seasons, say R_{2023} and R_{2021}, then we compare the rates by use of the difference d_{logit} of the corresponding logits:

$$d_{logit} = \log\left(\frac{R_{2023}}{100 - R_{2023}}\right) - \log\left(\frac{R_{2021}}{100 - R_{2021}}\right)$$

If the difference d_{logit} is positive (negative), this indicates that the 2023 rate is higher (lower) than the 2021 rate. We write a special function `logit()` to compute a logit. The `pivot_wider()` function helps us reorganize the data so that we can easily subtract the ball in play rates across years. Then by use of `mutate()`, we define the variable `d_BIP_logits` to be the difference of the

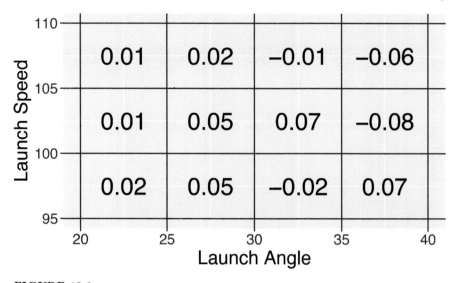

FIGURE 13.9
Difference in logit balls in play rates over regions for 2023 and 2021 seasons.

logits of the two balls-in-play rates. We use the `bin_plot()` in Figure 13.9 to display the values of `d_BIP_logits` over the launch variable space.

```
logit <- function(x){
  log(x) - log(100 - x)
}

S2 |>
  select(year, LA, LS, BIP_Rate) |>
  pivot_wider(
    names_from = year, names_prefix = "y",
    values_from = BIP_Rate
  ) |>
  mutate(d_BIP_logits = logit(y2023) - logit(y2021)) |>
  bin_plot(
    LA_breaks, LS_breaks,
    label = round(d_BIP_logits, 2)
  )
```

Note that most of the difference in logit values across the regions are positive, especially for exit velocities between 100 and 105 mph and launch angle values between 20 and 35 degrees. The takeaway is the 2023 rates are higher—the 2023 hitters are more likely than the 2021 hitters to put more balls in play in launch variable regions where home runs are likely.

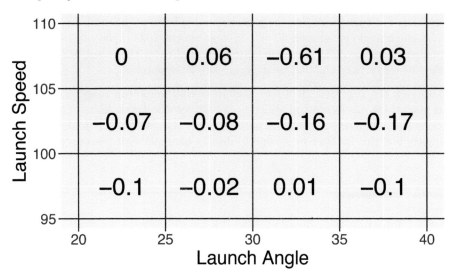

FIGURE 13.10
Difference in logit home run rates over regions for 2023 and 2021 seasons.

13.3.5   Changes in carry of the baseball

The last section focused on the behavior of the batters—they are hitting at higher rates of hard-hit balls with high launch variables. A second question relates to the carry properties of the baseball. Given values of the launch variables, what is the chance the batted ball is a home run?

We compare the home run rates in bins for the two seasons using the logit measure. We define the measure d_HR_logits which is the difference between the logit home run rate for 2023 and the corresponding rate for 2021. Again we use the bin_plot() function to display these differences in logits in Figure 13.10.

```
S2 |>
  select(year, LA, LS, HR_Rate) |>
  pivot_wider(
    names_from = year, names_prefix = "y",
    values_from = HR_Rate
  ) |>
  mutate(d_HR_logits = logit(y2023) - logit(y2021)) |>
  bin_plot(
    LA_breaks, LS_breaks,
    label = round(d_HR_logits, 2)
  )
```

Here we see that most of the difference values are negative which indicates that the logit of the HR rate for 2023 is smaller than the corresponding logit for 2021. For example, when the launch speed is between 100 and 105 mph and the launch angle is between 30 and 35 degrees, the 2023 home run rate is 0.16 smaller than the 2021 home run rate. This indicates that the 2023 baseball has less carry than the ball used during the 2021 season.

Although the 2021 and 2023 have similar overall home run rates on balls put in play, there are interesting differences between the two seasons. There is an increasing tendency in 2023 for batters to hit the ball harder in high launch angles, but this is offset by a slightly deader baseball with less carry in 2023.

13.4 Further Reading

Major League Baseball commissioned an independent study to investigate the causes of the home run increase during the period 2015–2017. The MLB commission released two reports: Albert et al. (2018) and Albert et al. (2019). A more recent study based on 2022 data is found in Albert and Nathan (2022).

13.5 Exercises

1. Modeling Probability of Hit

a. Using a similar generalized additive model as in Section 13.2.2, fit a model to the logit of the probability of a hit as a smooth function of the launch angle and launch velocity.

b. Using this model, predict the probability that a ball hit at 20 degrees at 100 mph will be a base hit.

c. Construct a contour plot of the fitted probability of a hit over the region where the launch angle is between 15 and 40 degrees and the launch speed is between 80 and 110 mph.

2. How Does P(HR) Depends on Spray Angle?

Construct a smooth scatterplot of the home run variable HR as a function of the spray angle similar to what was done in Section 13.2.3. Use this plot to show that it is hardest to hit a home run hit to dead center field.

3. Optimal Launch Angle for a Single

Define a new variable S that is equal to 1 if a single is hit and 0 otherwise. Fit a generalized linear model where the logit of the probability of a single is a smooth function of the launch angle. By looking a predictions from this model, find the launch angle which maximizes the probability of a single.

4. Nonnested Model for Launch Angle and Launch Speed

a. In Section 13.2.7, we fit a non-nested random effects model for the probability that a batted ball is a home run. Using the `lmer()` function in the **lme4** package, explore how the launch angles depend on the batter and the pitcher. Find the random effects standard deviations.

b. Fit a similar non-nested random effects model using launch velocity as a response variable. Again find the random effects standard deviations.

c. Looking at the results from parts a and b, is the variation in launch angles attributable primarily to the hitter or the pitcher? What about launch velocities—what is the primary source of the variation?

5. Comparing Home Run Hitting for Two Seasons

In Section 13.3, batted ball and home run rates for the 2021 and 2023 seasons were compared using a particular choice of bins for launch angle and launch speed. By using a different set of breakpoints for the two variables, compare batted ball rates and home runs for the two seasons. Compare your findings with the conclusions stated in Section 13.3.

6. Ratios in Rates for Two Seasons

In Section 13.3, batted ball and home run rates for the 2021 and 2023 seasons were compared using logits. Repeat the exercise using rates instead of logits and use ratios of rates to compare the two seasons. Explain using a few sentences how batted ball rates and home run rates have changed between the two seasons.

14

Making a Scientific Presentation using Quarto

Quarto is a new authoring system for producing many different kinds of documents that include dynamic computation of R (or Python) code. In many ways, Quarto is the natural successor to R Markdown. The previous edition of this book was written in LaTeX, with the **knitr** package providing a mechanism for rendering the R code into LaTeX, which was then compiled into a PDF. However, this book is written in Quarto. This change enables us to simultaneously render the book into multiple formats: in this case, HTML for the website, and PDF for the printed edition.

As noted above, Quarto documents can be rendered to a variety of different formats, including HTML presentations powered by `reveal.js`, a popular JavaScript library. In this chapter, we illustrate how Quarto an be used to create a professional quality reproducible scientific presentation.

14.1 Introduction to Slides in Quarto

Like all Markdown documents, Quarto documents have a *YAML* header, followed by various sections of Markdown code. In the example below, the YAML header appears at the top of the document, enclosed by `---` on the first line and `---` on the 4th line. Various options can be specified in the YAML section. In this case, we specify the `author`, `title`, and (importantly) the `format`. Note that `revealjs` is nested within `format`, and that `incremental` is nested within `revealjs`. What we are saying is that we want the `incremental` property, which is available within the `revealjs` format, to be set to `true`. In Quarto, it possible to specify different options for multiple formats simultaneously.

```
---
title: "Home Run Hitting"
author: "Jim Albert"
format:
```

```
revealjs:
   incremental: true
---
```

Next, we create slides using Markdown section headers and regular Markdown content. For example, the following Markdown code creates a slide titled "Baseball" that contains a list of two players: Mickey Mantle and Aaron Judge.

```
## Baseball

- Mickey Mantle
- Aaron Judge
```

In Section 14.2, we create a complex set of slides, based on extensions of these simple principles.

14.2 Example: Jim's Presentation on Home Run Hitting

In this section, we will walk you through the creation of a scientific presentation constructed in Quarto and output as HTML using the reveal.js framework. The full presentation as Jim gave it is available at: https://bayesball.github.i o/homerun_talk/homeruns.html. A slightly modified version that is consistent with the content presented here can be viewed at: https://beanumber.github.i o/abdwr3e/revealjs/hr_pres.html

14.2.1 Sections of presentation

First, we decide to divide the presentation into three sections, one introducing general patterns of home run hitting in baseball history, another describing the findings of the 2017 MLB Home Run Committee (Albert et al. 2018), and finally a section describing recent changes in the pattern of home run hitting during the Statcast era.

We create separate slides with these section titles by use of level 1 headings (#):

```
# Introduction

# What is Causing the Increase in Home Rate Rates?

# Recent Exploration of Home Run Rates
```

14.2.2   Including R output

One attractive aspect of using Quarto in creating presentations is the ability to incorporate R code and output into a document. As an illustration, suppose we are interested in describing the dramatic change in home run hitting over Major League Baseball history by graphing the number of home runs hit per team per game against season. We use the **Lahman** database to retrieve these data.

In the Quarto document, we display the R code within chunks demarcated by the ```` ```{r} ```` and ```` ``` ```` symbols. The compiled document contains the **ggplot2** graph displayed below. By using the `echo: true` option, the R code is included in the compiled document. In the case of a presentation, we would generally use the `echo: false` option so the code would not be shown. You can see the full slide in Jim's presentation.

```{r}
#| echo: true
library(Lahman)
br <- Batting |>
  group_by(Year = yearID) |>
  summarize(HR = sum(HR))
ggplot(br, aes(Year, HR)) +
  geom_point(color = "black", size = 2) +
  geom_smooth(
    formula = "y ~ x", color = "blue",
    se = FALSE, method = "loess",
    span = 0.20, linewidth = 1.5
  ) +
  labs(x = "Season", y = "Avg HR") +
  theme(text = element_text(size = 22))
```

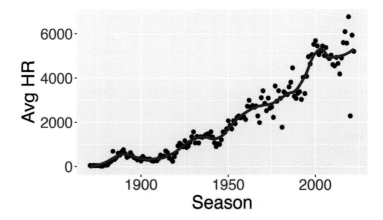

14.2.3 Multiple columns and adding images

Part of the presentation is describing some of the famous home run hitters in baseball history. The image file at the *URL* below contains a picture of Home Run Baker. Note that unlike Beamer presentations in LaTeX (see Section 14.3), images in `reveal.js` presentations can be sourced directly from the Internet. We include this image in the presentation by use of the `![](url)` syntax where `url` indicates the location of the folder containing the image.

The Quarto code also illustrates the use of a two column format where the left column contains the image and the right column gives a brief description of the player. The `.column width` argument describes the percentages for the widths of the left and right columns.

```
## Home Run Baker

:::: columns

::: {.column width="40%"}
![](https://live.staticflickr.com/6/12033666_c111eb7fab_z.jpg)
:::

::: {.column width="60%"}
-   Played during Deadball Era
-   Was home run leader in 1914 with 9 HR
-   Home runs were not a big part of the game
:::
::::
```

Figure 14.1 shows the completed slide of the compiled presentation.

14.2.4 Including a table

It is easy to include a table into the presentation by use of standard Markdown methods. Here we create a table displaying the total home run counts for the seasons 2015 through 2022 using the common Markdown pipe format. The resulting slide is shown in Figure 14.2.

```
## Home Run Totals in the Statcast Era

| Season | HR Total |
|--------|----------|
| 2015   | 4909     |
| 2016   | 5610     |
| 2017   | 6105     |
| 2018   | 5585     |
| 2019   | 6776     |
```

Home Run Baker

- Played during Deadball Era
- Was home run leader in 1914 with 9 HR
- Home runs were not a big part of the game

FIGURE 14.1
Slide from Jim's presentation showing Home Run Baker.

Home Run Totals in the Statcast Era

Season	HR Total
2015	4909
2016	5610
2017	6105
2018	5585
2019	6776
2021	5944
2022	5215

FIGURE 14.2
Slide from Jim's presentation showing a table of home runs counts.

```
| 2021   | 5944    |
| 2022   | 5215    |
```

An alternative method for displaying a table uses the `kable()` function in the **knitr** package. The table data is placed in a data frame by use of the `tibble()`

TABLE 14.1
A table of home runs counts rendered using `kable()`.

Season	Home runs
2015	4909
2016	5610
2017	6105
2018	5585
2019	6776
2021	5944
2022	5215

function. In the `kable()` function, we choose a "simple" format, align both columns to be centered, and add special names for the columns. Below we show the code to implement this R work and then display the output in Table 14.1.

```{r}
#| echo: false
df <- tibble(
  Season = c(2015, 2016, 2017, 2018, 2019, 2021, 2022),
  HR_Total = c(4909, 5610, 6105, 5585, 6776, 5944, 5215)
)
df |>
  knitr::kable(
    "simple", align = "cc",
    col.names = c("Season", "Home Runs")
  )
```

14.2.5 Including LaTeX

Another attractive feature of Quarto documents is the ability to incorporate LaTeX, which is a venerable system for displaying mathematical formulas. A LaTeX equation can be placed within the $$ delimiters. Here we use LaTeX to display the formula for the in-play home run rate. Below the Quarto code, Figure 14.3 shows a snapshot of the slide containing the mathematical expression.

```
## In-Play Rates

-   Define the home run rate as the fraction of $HR$ among all
    batted balls ($AB - SO$)
$$
HR \, Rate = \frac{HR}{AB - SO}
```

```
$$
-   Look at history of $HR$ rates
```

In-Play Rates

- Define the home run rate as the fraction of HR among all batted balls (AB − SO)

$$\text{HR Rate} = \frac{\text{HR}}{\text{AB} - \text{SO}}$$

- Look at history of HR rates

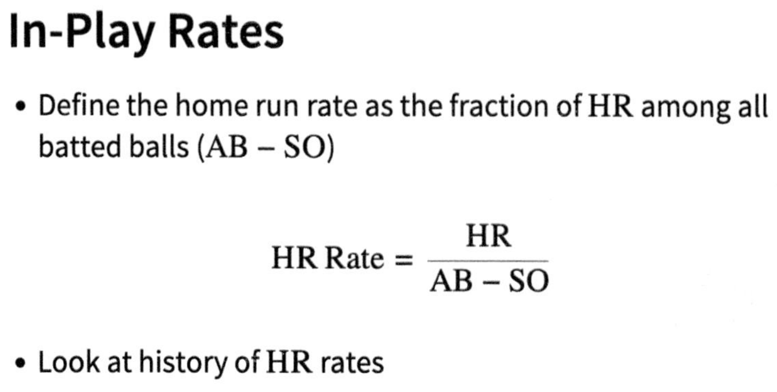

FIGURE 14.3
Slide from Jim's presentation showing an equation set in LaTeX.

14.2.6   Options using the `revealjs` format

The `revealjs` format (resulting in reveal.js presentations) allows for a wide variety of options. The website https://quarto.org/docs/presentations/revealjs/ provides an overview of all the capabilities of this particular presentation format.

There are many advanced features, including the ability to create animations, zoom in and out, include slides numbers, print to PDF, and a Chalkboard feature that allow you to draw on a slide.

There are 11 built-in themes for Reveal presentations—snapshots from the `default` theme have been presented up to this point. One can use the `sky` format by use of the following YAML header. A snapshot of the Home Run Baker slide is displayed using this theme in Figure 14.4.

```
---
title: "Home Run Hitting"
author: "Jim Albert"
format:
  revealjs:
    incremental: true
    theme: sky
---
```

HOME RUN BAKER

- Played during Deadball Era
- Was home run leader in 1914 with 9 HR
- Home runs were not a big part of the game

FIGURE 14.4
Slide from Jim's presentation showing Home Run Baker and using the sky theme.

14.3 Alternative Output Formats

We have focused on the use of the `revealjs` output format where the presentation output is an HTML file. Other popular formats are Beamer (where the output is a PDF file) and PowerPoint (where the output is a `.pptx` file).

These alternative formats require only small changes to the YAML header. For example, suppose one wishes to create a PDF of our home run presentation using the Beamer LaTeX class. In the YAML header, we indicate as an option to `format` that we wish to use the `beamer` class. In the options to `beamer`, we indicate we want to have incremental lists and use the `Boadilla` beamer theme with a `seahorse` color theme.

```
---
title: "Home Run Hitting"
author: Jim Albert
format:
  beamer:
    incremental: true
    theme: Boadilla
    colortheme: seahorse
---
```

Instead, suppose we wish to have PowerPoint output. We indicate below that the format is `pptx` with incremental lists. No other changes to the presentation content are necessary!

```
---
title: "Home Run Hitting"
author: Jim Albert
format:
  pptx:
    incremental: true
---
```

By default, PowerPoint will use a relatively plain looking template. You can modify this by creating a PowerPoint template and using the `reference-doc` option to use this template.

14.4   Further Reading

The official Quarto documentation provides an introduction to the HTML, PowerPoint and Beamer formats for creating presentations. The reveal.js site provides more details on the use of the reveal.js presentation framework.

14.5   Exercises

1. Home Run Hitting in Recent Years

Use the **Lahman** package and write R code that will generate the table of the total number of home runs hit from 2015–2022. Put the table on a slide as we have done in Section 14.2.4.

2. Home Run Hitting in Recent Years (continued)

Use the **Lahman** package and write R code that will display the home run rate on balls in play (`HR / (AB - SO)`) for seasons 2000–2022 as a function of the season. Put the figure on a slide as we have done in Section 14.2.2.

3. Should the Player be in the Hall of Fame?

Select a player who you believe should be in the Baseball Hall of Fame. Make a presentation that compares your player with another player at the same fielding position who is currently in the Hall of Frame. Your presentation should include graphs, tables and images of both players.

4. Leaderboard Presentation

Make a presentation that presents the top-ten career leaders with respect to some batting or pitching measure. This presentation should include an R scatterplot that displays the career measure against the midcareer season for the ten players. Also include images of the ten players in the leaderboard together with some information about the players.

15

Using Shiny for Baseball Applications

15.1 Introduction

One of the exciting features of the R ecosystem is the relative ease in constructing web applications of R work by use of the **shiny** package. In this chapter, we illustrate the construction of a Shiny app by use of a baseball application where one wishes to compare the career trajectories of two pitchers from Major League Baseball history.

A good starting point to develop a Shiny app is writing a function that implements the computation that one wishes to display in the app. In Section 15.2, we use several R functions to select a group of contemporary pitchers and construct the comparison graph of their career trajectories. Section 15.3 outlines the steps of constructing the Shiny app including the user interface and server components and running the app. Once the Shiny app is completed, Section 15.4 describes several methods of getting other people to try your app and Section 15.5 concludes by providing some tips that should help interested readers get their own apps running quickly.

15.2 Comparing Two Pitcher Trajectories

We focus on comparing the career trajectories of two pitchers who played during the same baseball era. Given a particular interval of seasons of interest and minimum number of innings pitched, we wish to graphically display a measure of performance against season or age for two selected pitchers. The relevant data is in the **Lahman** package and a FanGraphs table containing values needed in the computation of the FIP measure.

We wrote two functions to help with these tasks. The first is a function called `selectPlayers2()` that returns a data frame of all of the pitchers who achieved

DOI: 10.1201/9781032668239-15

a certain minimum number of innings pitched where the pitchers' midcareer fell inside a particular time interval. There are too many pitchers in the history of baseball to list them all—to do so would make the app cumbersome to use. Instead, `selectPlayers2()` helps us narrow the list to a reasonable number of pitchers. For example, the following code returns all of the pitchers with at least 2000 innings pitched and whose midcareer fell between 1959 and 1966. These are the pitchers who are eligible to be compared by the app.

```
library(abdwr3edata)
selectPlayers2(c(1959, 1966), 2000)
```

```
# A tibble: 26 x 2
   playerID  Name
   <chr>     <chr>
 1 bellga01  Gary Bell
 2 buhlbo01  Bob Buhl
 3 bunniji01 Jim Bunning
 4 cardwdo01 Don Cardwell
 5 chancde01 Dean Chance
 6 drysddo01 Don Drysdale
 7 ellswdi01 Dick Ellsworth
 8 grantmu01 Mudcat Grant
 9 jacksla01 Larry Jackson
10 klinero01 Ron Kline
# i 16 more rows
```

Inside the function itself, we begin by computing the mid-career year and number of innings pitched, where `midYear` is defined to be the average of the first and final seasons of a pitcher. Given the aforementioned values as inputs, `selectPlayers2()` queries the **Lahman** package and outputs the player ids and names for all pitchers that meet the criteria. The full code for the function is shown below.

```
selectPlayers2
```

```
function(midYearRange, minIP) {
  Lahman::Pitching |>
    mutate(IP = IPouts / 3) |>
    group_by(playerID) |>
    summarize(
      minYear = min(yearID),
      maxYear = max(yearID),
      midYear = (minYear + maxYear) / 2,
      IP = sum(IP),
```

```
      .groups = "drop"
    ) |>
    filter(
      midYear <= max(midYearRange),
      midYear >= min(midYearRange),
      IP >= minIP
    ) |>
    select(playerID) |>
    inner_join(Lahman::People, by = "playerID") |>
    mutate(Name = paste(nameFirst, nameLast)) |>
    select(playerID, Name)
}
<bytecode: 0x59bb9e639d20>
<environment: namespace:abdwr3edata>
```

A second helper function `compare_plot()` constructs the graph comparing the career trajectories of two selected pitchers. This function requires the player ids for the two pitchers, the measure to graph (among ERA, WHIP, FIP, SO Rate, BB Rate) on the vertical axis, and the time variable (either `season` or `age`) to plot on the horizontal axis.

To illustrate the use of the `compare_plot()` function, suppose we wish to compare the FIP (fielding-independent pitching) trajectories as a function of age for the great Dodgers pitches Sandy Koufax and Don Drysdale. From the **People** table in the **Lahman** package, we collect the player ids for the two pitchers. The `fg` data frame contains data from the FanGraphs "guts" table. We then apply the `compare_plot()` function with inputs `koufasa01`, `drysddo01`, FIP, and `age` (see Figure 15.1).

```
compare_plot(
  "koufasa01", "drysddo01", "FIP", "age"
) |>
  pluck("plot1")
```

For each pitcher, this function constructs a scatterplot of the FIP measure against age and overlays a smoothing curve. The `geom_textsmooth()` function from the **geomtextpath** package is used to add player labels to each smoothing curve. The full code for the function is a bit long to display here, but you can access the code from the **abdwr3edata** package.

```
compare_plot
```

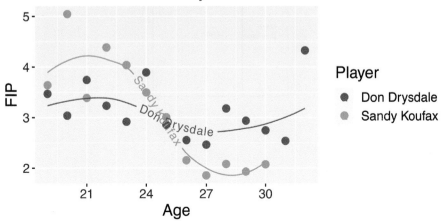

FIGURE 15.1
Career trajctories of FIP for Sandy Koufax and Don Drysdale.

15.3 Creating the Shiny App

15.3.1 Basic structure

A Shiny app is contained in a single R script file frequently named `app.R`. This file contains three basic components:

- a user interface object `ui` describing the layout of the app include all input controls
- a server function `server()` describing the instructions needed to run the app
- a call to the `shinyApp()` function creating the app given the user interface and server information

The following code displays the basic structure of the `app.R` file. Note that this file initially lists the two functions `selectPlayers2()` and `compare_plot()` followed by the Shiny component `ui` and the Shiny functions `server()` and `shinyApp()`.

```
library(shiny)

selectPlayers2 <- function(midYearRange, minIP) {
  # ...code...
```

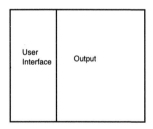

FIGURE 15.2
Layout of one Shiny app.

```
}

compare_plot <- function(playerid_1, playerid_2, measure, xvar, fg) {
  # ...code...
}

ui <- fluidPage(
  # ...code...
)

server <- function(input, output, session) {
  # ...code...
}

shinyApp(ui = ui, server = server)
```

You can view the full code for the app through the `compareTrajectories()` function from the **abdwr3edata** package.

```
compareTrajectories
```

15.3.2   Designing the user interface

In the layout of this particular Shiny app, the user interface controls are on the left side of the app and the output is on the right side as shown in Figure 15.2.

The layout is defined by use of the `fluidPage()` function inside the `ui` object. The `fluidRow()` function defines a Shiny output window that is 4 units wide for the user interface and 8 units wide for the output.

```
ui <- fluidPage(
  fluidRow(
    column(4,
      # user interface controls
    ),
    column(8,
      # output functions
    )
  )
)
```

The user interface controls for this application consist of sliders, pull-down menus and radio buttons. Functions from the **shiny** package are used to construct the different input types in the app.

Slider controls are used to input the range of mid-career and minimum innings pitched (IP) values. The `sliderInput()` function is used to define the first slider input `midyear`. The inputs to this function are the input label, the text to display, the range of slider values, and the current value. Since `value` is a vector of two values, one is inputting a range of values in the slider.

```
sliderInput(
  "midyear",
  label = "Select Range of Mid Career:",
  min = 1900, max = 2010,
  value = c(1975, 1985),
  sep = ""
)
```

A `selectInput()` function is used to construct a pull-down menu input item. We display below the code for inputting the `player_name1` variable. Note that the `selectPlayers2()` function is used to produce the list of player names that have specific midcareer and minimum PA values.

```
selectInput(
  "player_name1",
  label = "Select First Pitcher:",
  choices = selectPlayers2(c(1975, 1985), 2000)$Name
)
```

Radio buttons are defined by use of the `radioButtons()` function. Below the code is displayed for the `type` variable. The inputs to this function are the label, the string that is displayed and the possible input values.

```
radioButtons(
  "type",
  label = "Select Measure:",
  choices = c("ERA", "WHIP", "FIP", "SO Rate", "BB Rate")
)
```

15.3.3   Adding dynamic user inputs

One special feature of this particular Shiny app is the use of dynamic UI where the values of the input controls can be modified by other input controls. Dynamic UI is achieved by use of the `observeEvent()` function in the `server()` function. In the following code snippet, in `observeEvent()`, the values of the `player_name1` input are modified when values of the `midyear` input are changed. The `observeEvent()` function is used several times so that the values of `player_name1` and `player_name2` are modified whenever the values of `midyear` or `midpa` are changed.

```
observeEvent(
  input$midyear,
  updateSelectInput(
    inputId = "player_name1",
    choices = selectPlayers2(
      input$midyear, input$minpa
    )$Name
  )
)
```

15.3.4   Completing the server component

The `server()` function also contains the actual work for the Shiny app. The following snippet shows how the output component `output$plot1` is defined. From the user inputs `input$midyear`, `input$midpa`, `input$player_name1` and `input$player_name2`, the `selectPlayers2()` function is applied to access the player user ids. Then the `compare_plot()` function is used with these inputs to construct the plot. The `renderPlot()` function controls how what is drawn on the app changes when these inputs change.

```
output$plot1 <- renderPlot({
  S <- selectPlayers2(input$midyear, input$minpa)
  id1 <- filter(S, Name == input$player_name1)$playerID
  id2 <- filter(S, Name == input$player_name2)$playerID
  compare_plot(id1, id2, input$type, input$xvar)$plot1
},
  res = 96
)
```

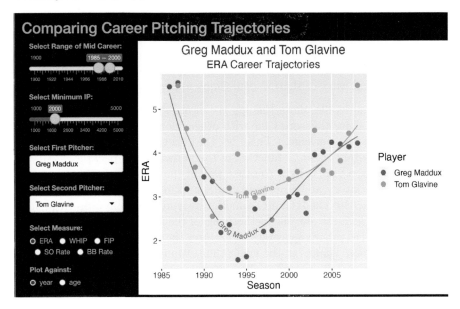

FIGURE 15.3
Snapshot of the career trajectories Shiny app.

15.3.5 Running the app

In usual practice, the `app.R` script containing the Shiny code is placed in a separate folder. One runs the Shiny app by typing in the RStudio console window

```
runApp()
```

Alternatively, one can press the "Run App" button at the top of the screen. Figure 15.3 displays a snapshot of the completed Shiny app. Because this particular app is part of an R package, you can run the app by typing:

```
compareTrajectories()
```

In Figure 15.3, one is selecting the mid-career interval 1985–2000, a minimum PA value of 2000, and comparing the ERA trajectories of the Hall of Fame pitchers Greg Maddux and Tom Glavine.

We note that while their career trajectories were similar, Maddux had a superior ERA during his peak years.

15.4   Sharing the App

There are several ways of sharing your Shiny app with others.

- **Share the app.R file.** Since the app is contained in a single file **app.R**, one can simply share this script file with other people.
- **Put it in a package.** This is the method illustrated by the `compareTrajectories()` function in the **abdwr3edata** package.
- **Share the app via Github.** Another way of sharing the app is to create a Github repository and store your Shiny app in that repository. Then the user can use the `runGitHub()` function to run the Shiny app from your repository. To illustrate this method, one of the authors created the Github repository **testshinyapp** and then stored the career trajectories app in this repository. Thanks to the `runGitHub()` function, the interested reader can run this app by typing in the Console.

```
runGitHub( "testshinyapp", "bayesball")
```

- **Host the app on a Shiny server.** Posit currently has a hosting service which allows a user to see your app as a web program. To use the Posit service, one needs to set up an account on https://www.shinyapps.io/.
  Then once you have your Shiny app running, there is a Publish button on the app display that uploads your app to the server. One of the authors recently did this for the career trajectory Shiny app and the live version of the app is current available at the following URL:

https://bayesball.shinyapps.io/CareerTrajectoryPitching/

15.5   Tips for Getting Started Making Apps

An easy way to get started is to start with a template, a script to a Shiny app that has a similar user interface to the one that you are interested to making. For example, if you are interested in plotting career trajectories for batters, you can modify the `CareerTrajectoryPitching` app described in this chapter.

There are many illustrations of the code for producing different types of Shiny apps on the Posit Shiny Gallery. By starting with a sample **app.R** script, one can avoid the small coding errors that are easy to make when one is constructing a program from scratch.

15.6 Further Reading

Posit has a large amount of information and examples of Shiny apps at the Shiny R site https://shiny.posit.co/r/getstarted/shiny-basics/lesson1/index.html. In addition, one of the authors has created an R package **ShinyBaseball** found at https://github.com/bayesball/ShinyBaseball that contains a large number of Shiny apps for illustrating baseball research for a variety of problems. These apps have been used to illustrate R work for the "Exploring Baseball with R" blog at https://baseballwithr.wordpress.com/.

15.7 Exercises

1. Plotting Locations of Balls in Play

The following function `construct_zone_plot()` produces a plot of the zone locations of balls in play for a player where the color of the plotting plot depends on the outcome. The inputs to the function are the Statcast dataset of balls in play `sc_ip`, the name of the batter `p_name` and the outcome `type` (either "Hit" or "Home_Run"). For example, if `sc2023_ip` is a data frame of balls in play for the 2023 season, then one can display the locations of all of Ronald Acuña's balls in play colored by hit by use of the function

```
construct_zone_plot(sc2023_ip, "Acuña Jr., Ronald", "Hit")
```

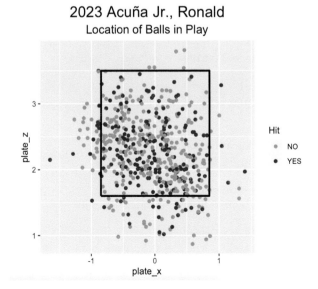

Construct a Shiny app using this function where the player name is input through a select list and the outcome type is input using radio buttons.

```
construct_zone_plot <- function(sc_ip, p_name, type) {
  require(dplyr)
  require(ggplot2)
  add_zone <- function() {
    topKzone <- 3.5
    botKzone <- 1.6
    inKzone <- -0.85
    outKzone <- 0.85
    kZone <- data.frame(
      x = c(inKzone, inKzone, outKzone, outKzone, inKzone),
      y = c(botKzone, topKzone, topKzone, botKzone, botKzone)
    )
    geom_path(aes(.data$x, .data$y),
      data = kZone, lwd = 1
    )
  }
  hits <- c("single", "double", "triple", "home_run")
  sc_player <- filter(sc_ip, player_name == p_name) |>
    mutate(
      Hit = ifelse(events %in% hits, "YES", "NO"),
      Home_Run = ifelse(events == "home_run", "YES", "NO")
    )
  ggplot() +
    geom_point(
      data = sc_player,
      aes(plate_x, plate_z,
        color = .data[[type]]
      )
    ) +
    add_zone() +
    coord_equal() +
    scale_colour_manual(values = c("tan", "red")) +
    labs(
      title = paste(
        substr(sc_player$game_date[1], 1, 4),
        p_name
      ),
      subtitle = "Location of Balls in Play"
    ) +
    theme(
      plot.title = element_text(
        color = "black", hjust = 0.5, size = 18
      ),
      plot.subtitle = element_text(
        color = "black", hjust = 0.5, size = 14
```

```
        )
      )
  }
```

2. Plotting Locations of Balls in Play using Other Outcomes

In Exercise 1, the color of the plotting point can depend on the outcome "Hit"
or "Home_Run". Revise the `construct_zone_plot()` function so that the `type`
outcome can be one of the continuous variables `launch_angle`, `launch_speed`
or `estimated_ba_using_speedangle`. Revise the Shiny app so that the user
can input one of these three variables.

3. Plotting Locations of Balls in Play with Brushing

Shiny allows one to interactively select portions of a graph by brushing. The
following code in the user input section modifies the `plotOutput()` function
by adding the `brush` option.

```
plotOutput("plot", brush = brushOpts("plot_brush", fill = "#0000ff"))
```

In a new `output$data` component of the server section of the Shiny app, the
following code will take a subset of the `sc_player` data frame which is defined
by the selected rectangle that is brushed.

```
sc1 <- brushedPoints(sc_player, input$plot_brush)
```

By using this code, modify the Shiny app in Exercise 1 to allow brushing of
the scatterplot. In a separate region of the display, compute the balls in play,
the hits, the home runs, and the corresponding hit and home run rates for
points in the selected region.

Appendices

A

Retrosheet Files Reference

A.1 Downloading Play-by-Play Files

A.1.1 Introduction

The play-by-play data files for every Major League season between 1913 and 2022 are currently available at the Retrosheet Web page at https://www.retr osheet.org/game.htm. By clicking on a single year, say 1950, one obtains a compressed (.zip) file containing a collection of files: one set of files containing information on the plays for the home games for all teams, and another set of files giving the rosters of the players for each team. This appendix illustrates the easiest way to work with Retrosheet files.

A.1.2 Chadwick

Henry Chadwick was a sportswriter who is credited with inventing the box score, batting average, and earned run average. The special software tools designed to process Retrosheet data are named in his honor. These tools are maintained by Ted Turocy, and are available at https://github.com/chadwic kbureau/chadwick. Please follow the installation instructions for Chadwick. The repository contains binaries suitable for Windows users, while Linux and Mac users can compile their own versions of these tools by downloading and compiling the source code.

The particular component of Chadwick that we need to generate Retrosheet play-by-play data is called cwevent. It is a program that is run at the command line. If it is installed and working properly, you can simply type cwevent at the command line and see output like this.

```
cwevent

Chadwick expanded event descriptor, version 0.10.0
  Type 'cwevent -h' for help.
Copyright (c) 2002-2023
Dr T L Turocy, Chadwick Baseball Bureau (ted.turocy@gmail.com)
```

This is free software, subject to the terms of the GNU GPL
license.

If you have installed Chadwick and you get an error when you run `cwevent`, it is
probably one of two problems, both of which involve setting path *environment
variables*. If the error says `command not found` (or something of that nature),
then your operating system cannot find the `cwevent` binary, probably because
your `PATH` environment variable does not include the directory where the
`cwevent` binary is. On this Ubuntu machine[1], we can find the correct path by
typing:

```
which cwevent
```

/usr/local/bin

You can check the current value of the `PATH` environment variable at the
command line using `echo`.

```
echo $PATH
```

Use the `export` directive to append the path to the `cwevent` binary to the
current `PATH` environment variable.

```
export PATH=$PATH:/usr/local/bin
```

If `cwevent` runs, but throws an error, the most likely problem is that `cwevent`
can't find the Chadwick shared libraries. You can solve this problem by setting
the `LD_LIBRARY_PATH` environment variable. Note that environment variables
are system-specific. On this Ubuntu machine, we can find the Chadwick shared
libraries using the `find` command.

```
find /usr/local -name "libchadwick*"
```

/usr/local/lib/libchadwick.la
/usr/local/lib/libchadwick.a
/usr/local/lib/libchadwick.so
/usr/local/lib/libchadwick.so.0
/usr/local/lib/libchadwick.so.0.0.0

So in order for `cwevent` to work, the `LD_LIBRARY_PATH` environment variable needs to include `/usr/local/lib`. You can set the `LD_LIBRARY_PATH`
environment variable using `export` as we did above, or, to set environment variables from within R, use the `Sys.getenv()` and `Sys.setenv()`

---

[1] On Windows, the analogous DOS command is `where`.

functions. The `safe_add_ld_path()` function we present below is similar to the `chadwick_ld_library_path()` function in the **baseballr** package.

```
safe_add_ld_path <- function(path_new = "/usr/local/lib") {
  path_old <- Sys.getenv("LD_LIBRARY_PATH")
  path_old_parts <- path_old |>
    str_split_1(":") |>
    unique()
  if (!path_new %in% path_old_parts) {
    path_new_parts <- c(path_new, path_old_parts)
    Sys.setenv(
      LD_LIBRARY_PATH = paste0(path_new_parts, collapse = ":")
    )
  }
  Sys.getenv("LD_LIBRARY_PATH")
}
```

```
safe_add_ld_path()
```

```
[1] "/usr/lib/R/lib:/usr/lib/x86_64-linux-gnu:/usr/local/lib:"
```

A.1.3   Downloading data for one or more seasons

Once you have Chadwick installed and working properly, the `retrosheet_data()` function from the **baseballr** package makes obtaining Retrosheet play-by-play data a breeze.

The `retrosheet_data()` function takes three optional arguments that help you manage your data. Note that this function calls **cwevent**, and so it will not work if you haven't set up Chadwick properly as in Section A.1.2.

To download the Retrosheet data we use in this book, type:

```
retro_data <- baseballr::retrosheet_data(
  here::here("data_large/retrosheet"),
  c(1992, 1996, 1998, 2016)
)
```

This will download and process four years worth of play-by-play data and return a `list` of length four, with each item containing a `list` of length two. Each of the four items in `retro_data` corresponds to the four years we specified. The two items in each of those years are data frames: one called `events` that stores the play-by-play data, and another called `rosters` that stores the rosters.

To isolate the play-by-play data for a single year, use the `pluck()` function.

```
retro1992 <- retro_data |>
  pluck("1992") |>
  pluck("events")
```

A.1.4   Saving the data

You probably don't want to have to download and process all of this data every time you want to use it. Now that you have it stored in R, the best way to save it for later use is by writing the data frame for each year to disk using R's internal data storage format. You can do this with the `write_rds()` function.

```
retro1992 |>
  write_rds(
    file = here::here("data/retro1992.rds"),
    compress = "xz"
  )
```

Be sure to use the `compress` argument—it will significantly reduce the size of the data.

If you want to build a whole database of Retrosheet data (for many years), iterate the process described above and combine it with our illustration of how to build a SQL database in Chapter 11 and our discussion of working with large data in Chapter 12.

A.1.5   The function `parse_retrosheet_pbp()`

In previous versions of this book, we included a function called `parse_retrosheet_pbp()` that could be used to download and process Retrosheet data. This function has been superseded by the `retrosheet_data()` function from the **baseballr** package, and we no longer recommend using it. However, if you are interested in working through its logic, the code is available through the **abdwr3edata** package.

```
abdwr3edata::parse_retrosheet_pbp
```

```
function(season) {
  download_retrosheet(season)
  unzip_retrosheet(season)
  create_csv_file(season)
  create_csv_roster(season)
  cleanup()
```

```
}
<bytecode: 0x575fc1168c20>
<environment: namespace:abdwr3edata>
```

You may want to access the various helper functions for further detail. For example:

```
abdwr3edata::download_retrosheet
```

```
function(season) {
    # get zip file from retrosheet website
    utils::download.file(
        url = paste0(
            "http://www.retrosheet.org/events/", season, "eve.zip"
        ),
        destfile = file.path(
            "retrosheet", "zipped",
            paste0(season, "eve.zip")
        )
    )
}
<bytecode: 0x575fc1330600>
<environment: namespace:abdwr3edata>
```

A.1.6   Alternatives to Chadwick

The **retrosheet** package (Douglas and Scriven 2024) provides an alternative method for bringing Retrosheet data into R without an external dependency on the Chadwick software through its `getRetrosheet()` function. However, these data are returned as a list of lists (rather than a list of data frames), and thus can be considerably more cumbersome to analyze.

A.2   Retrosheet Event Files: a Short Reference

As we mentioned in Chapter 1, Retrosheet event files come in a format expressly devised for them, and require the use of some software tools for converting them in a format suitable for data analysis. Retrosheet provides such software tools https://www.retrosheet.org/tools.htm and a step-by-step example https://www.retrosheet.org/stepex.txt for performing the conversion.

Chadwick provides similar tools for parsing Retrosheet event files that have been used for creating the play-by-play files used in this book (see Section A.1.2).

Chadwick tools generate a line for each play in the Retrosheet event files, consisting of 97 "regular" columns (the same that are obtained using the tools provided by Retrosheet) plus 63 "extended" fields, allowing to easily access all of the information contained in the Retrosheet event files. Going through every one of the more than 150 columns generated by the Chadwick tools is beyond the scope of this book, and thus we point to the documentation on the Chadwick website for the full list.[2] In this section we present the main fields describing an event and the state of the game when it happens.

Please note that while the Chadwick tools return variable names in all caps, the `retrosheet_data()` function uses *snake_case* for variable names.

A.2.1 Game and event identifiers

The games are identified in Retrosheet event files by 12-character strings (the `GAME_ID` column): the first three characters identify the home team, the following eight characters indicate the date when the game took place (in the YYYYMMDD format), and the last character is used to distinguish games of doubleheaders (thus "1" indicates the first game, "2" the second game, and "0" means only one game was played on the day).

Events are progressively numerated in each game (column `EVENT_ID`), thus every single action in the Retrosheet database can be uniquely identified by the combination of the game identifier and the event identifier.

A.2.2 The state of the game

Several fields are helpful for defining the state of the game when a particular event happened. The **inning** and the **team on offense** variables are stored in the `INN_CT` and `BAT_HOME_ID` fields respectively. The latter field can assume values "0" (away team batting, i.e., top of the inning) or "1" (home team batting, bottom of the inning). The **visitor score** and the **home score** variables are recorded in the `AWAY_SCORE_CT` and `HOME_SCORE_CT`.

The **number of outs** before the play is indicated in the `OUTS_CT` column, while the situation of **runners on base** is coded in the field `START_BASES_CD`, using numbers from 1 to 7 as shown in Table A.1.[3]

The actual **description of the event** resides in the `EVENT_TX` column, consisting of a string describing the outcome of the play (e.g., strikeout, single, etc.), some additional details (e.g., the type and location of the batted ball), and the advancement of any runner on base. Several columns are generated by

---

[2]The documentation for all the software tools is available at https://chadwick.sourceforge.net/doc/cwtools.html. In particular, the tool for processing the event files (`cwevent`) is documented at https://chadwick.sourceforge.net/doc/cwevent.html#cwtools-cwevent.

[3]An analogous column named `END_BASES_CD` contains the base state at the end of the play, coded in the same way.

TABLE A.1

Retrosheet coding for the situation of runners on base.

Code	Bases occupancy
0	Empty
1	1B only
2	2B only
3	1B & 2B
4	3B only
5	1B & 3B
6	2B & 3B
7	Loaded

decoding the EVENT_TX string:

- EVENT_CD is a numeric code reflecting the **basic event**; Table A.2 displays the codes for the possible plays coded in this column.

- BAT_EVENT_FL is a flag indicating whether an event is a **batting event**, in which case it is labeled as T. Non-batting events include, for example, stolen bases, wild pitches and, generally, any event that does not mark the end of a plate appearance.

- H_CD is a numeric code indicating the **base hit type**, going from 1 for a single to 4 for a home run.

- BATTEDBALL_CD is a single character code denoting the **batted ball type**. It can assume one of the following values: G (ground ball), L (line drive), F (fly ball), P (pop-up). Note that for most of the seasons in the Retrosheet database, the batted ball type is reported only for plate appearances ending with the batter making an out, while they are not available on base hits.

- BATTEDBALL_LOC_TX is a string indicating the **batted ball location**, coded according to the diagram shown at https://www.retrosheet.org/location.htm. Note that this information is available for a limited number of seasons.

- FLD_CD is a numeric code denoting the **fielder** first touching a batted ball, coded with the conventional baseball fielding notation going from 1 (the pitcher) to 9 (the right fielder).

The **sequence of pitches** is recorded in the PITCH_SEQ_TX and has been addressed in Chapter 6, where Table 6.1 displays how the different pitch outcomes are coded. Several columns are generated from this one, indicating counts of the various types of pitch outcomes, as displayed in Table A.3.

TABLE A.2

Retrosheet coding for the type of event.

Code	Event type
2	Generic Out
3	Strikeout
4	Stolen Base
5	Defensive Indifference
6	Caught Stealing
8	Pickoff ·
9	Wild Pitch
10	Passed Ball
11	Balk
12	Other Advance
13	Foul Error
14	Nonintentional Walk
15	Intentional Walk
16	Hit By Pitch
17	Interference
18	Error
19	Fielder Choice
20	Single
21	Double
22	Triple
23	Homerun

TABLE A.3

Columns reporting counts of various pitch types.

Column name	Column description
PA_BALL_CT	No. of balls in plate appearance
PA_CALLED_BALL_CT	No. of called balls in plate appearance
PA_INTENT_BALL_CT	No. of intentional balls in plate appearance
PA_PITCHOUT_BALL_CT	No. of pitchouts in plate appearance
PA_HITBATTER_BALL_CT	No. of pitches hitting batter in plate appearance
PA_OTHER_BALL_CT	No. of other balls in plate appearance
PA_STRIKE_CT	No. of strikes in plate appearance
PA_CALLED_STRIKE_CT	No. of called strikes in plate appearance
PA_SWINGMISS_STRIKE_CT	No. of swinging strikes in plate appearance
PA_FOUL_STRIKE_CT	No. of foul balls in plate appearance
PA_INPLAY_STRIKE_CT	No. of balls in play in plate appearance
PA_OTHER_STRIKE_CT	No. of other strikes in plate appearance

A.3 Parsing Retrosheet Pitch Sequences

A.3.1 Introduction

Chapter 6 showed how to compute, using regular expressions, whether a plate appearance went through either a 1-0 or a 0-1 count. Here we provide the code to retrieve the same information for every possible balls/strikes count.

A.3.2 Setup

We first load Retrosheet data for the 2016 season.

```
retro2016 <- read_rds(here::here("data/retro2016.rds"))
```

Then a new column **sequence** is created in which the pitch sequence is reported, stripped by any character not indicating an actual pitch to the batter.[4]

```
retro2016 <- retro2016 |>
  mutate(sequence = gsub("[.>123+*N]", "", pitch_seq_tx))
```

A.3.3 Evaluating every count

Every plate appearance starts with a 0-0 count. The code for both the 1-0 and 0-1 counts was described in Chapter 6.

```
retro2016 <- retro2016 |>
  mutate(
    c00 = TRUE,
    c10 = grepl("^[BIPV]", sequence),
    c01 = grepl("^[CFKLMOQRST]", sequence)
  )
```

A number inside the curly brackets indicates the exact number of times the preceding expression has to be repeated in the string to match. The following lines look for plate appearances going through the counts 2-0, 3-0, and 0-2.

```
retro2016 <- retro2016 |>
  mutate(
    c20 = grepl("^[BIPV]{2}", sequence),
    c30 = grepl("^[BIPV]{3}", sequence),
```

---

[4]See Table 6.1 in Chapter 6 for reference.

```
    c02 = grepl("^[CFKLMOQRST]{2}", sequence)
)
```

The | (vertical bar) character is used to separate alternatives. The following lines parse the `sequence` string looking for the different sequences that can lead to 1-1, 2-1, and 3-1 counts.

```
b <- "[BIPV]"
s <- "[CFKLMOQRST]"
retro2016 <- retro2016 |>
  mutate(
    c11 = grepl(
      paste0("^", s, b, "|^", b, s), sequence
    ),
    c21 = grepl(
      paste0("^", s, b, b,
             "|^", b, s, b,
             "|^", b, b, s), sequence
    ),
    c31 = grepl(
      paste0("^", s, b, b, b,
             "|^", b, s, b, b,
             "|^", b, b, s, b,
             "|^", b, b, b, s), sequence
    )
  )
```

On two-strike counts, batters can indefinitely foul pitches off without affecting the count. In the lines below, sequences reaching two strikes before reaching the desired number of balls feature the `[FR]*` expression, denoting a foul ball[5] happening any number of times, including zero, as indicated by the asterisk.

```
retro2016 <- retro2016 |>
  mutate(
    c12 = grepl(
      paste0("^", b, s, s,
             "|^", s, b, s,
             "|^", s, s, "[FR]*", b), sequence
    ),
    c22 = grepl(
      paste0("^", b, b, s, s,
             "|^", b, s, b, s,
             "|^", b, s, s, "[FR]*", b,
             "|^", s, b, b, s,
```

---
[5]F encodes a foul ball, R a foul ball on a pitchout. See Table 6.1.

```
            "|^", s, b, s, "[FR]*", b,
            "|^", s, s, "[FR]*", b, "[FR]*", b),
    sequence
  ),
  c32 = grepl(
    paste0("^", s, "*", b, s,
           "*", b, s, "*", b), sequence
  ) & grepl(
    paste0("^", b, "*", s, b, "*", s), sequence
  )
)
```

The `retrosheet_add_counts()` in the **abdwr3edata** package contains all of the code necessary to compute these counts.

```
abdwr3edata::retrosheet_add_counts
```

B

Historical Notes on PITCHf/x Data

B.1 Introduction

PITCHf/x was a product by Sportvision, a company that produced broadcast effects for sports, such as the first-down virtual line for football and the FoxTrax hockey puck. Two cameras installed in each MLB park recorded the flight of the baseball between the pitcher's mound and home plate, and advanced software calculated the position, the velocity, and the acceleration of the ball, giving sufficient information to estimate the full trajectory of the ball in its mound-to-plate trip.

From 2006 until fairly recently, Major League Baseball Advanced Media (MLBAM) maintained a publicly accessible Gameday web server that fed their online content with real-time data delivered in an XML format. Previous editions of this book used the **pitchRx** package to download this pitch-by-pitch information. Unfortunately, this server has been superseded by an *API*-based system located at https://statsapi.mlb.com/, and thus the **pitchRx** package no longer works. While this new system is also publicly accessible and delivers real-time data, it is not well documented and proper usage requires registration as a developer. Several developers have published API packages on GitHub, most notably the `MLB-StatsAPI` package for Python available at https://github.com/toddrob99/MLB-StatsAPI, but to the best of our knowledge no similar R package is in widespread use.

That is the bad news.

The good news is that most, if not all, of the useful data provided by the old Gameday server is available through the Statcast data served via Baseball Savant. Appendix C details the retrieval and use of these data.

In the remainder of this section, we list several resources that were vital in moving sabermetric research fueled by PITCHf/x data forward, mainly for posterity.

DOI: 10.1201/9781032668239-B

B.2   Online Resources

The following PITCHf/x resources were available on the World Wide Web. Note that due to site maintainers being hired by MLB front offices, the demise of PITCHf/x, or exclusive licensing contracts, these resources are subject to being removed or moved.

- **Mike Fast's PITCHf/x glossary** (https://fastballs.wordpress.com/2007/08/02/glossary-of-the-gameday-pitch-fields/): Detailed explanations for the PITCHf/x fields have been provided by Mike Fast.

- **Brooks Baseball** (https://www.brooksbaseball.net/): Created and maintained by Dan Brooks, its main features are the *Player Cards*, consisting of tables and charts for every pitcher who has ever played in a ballpark with the PITCHf/x system installed. Tables and charts report information on characteristics of pitches, their usage (including sequencing), and the outcomes they produce. The classification of pitches used at Brooks Baseball is not the MLBAM one, as pitches are classified by *Pitch Info LLC*. Another useful resource of Brooks Baseball is the *PitchFX Tool*, which allows site visitors to select one pitcher for one game and obtain a pitch-by-pitch table.

- **Baseball Prospectus** (https://www.baseballprospectus.com/): In its *Statistics* section, Baseball Prospectus offers *PITCHf/x Hitters Profiles*, *PITCHf/x Pitchers Profiles*, *PITCHf/x Leaderboards*, and *PITCHf/x Matchups*. The building blocks of these resources come from the previously mentioned Brooks Baseball.

- **FanGraphs** (https://www.fangraphs.com/): FanGraphs has PITCHf/x tables and charts for individual players. For example, pitcher James Shields's PITCHf/x page is available at https://www.fangraphs.com/pitchfx.aspx?playerid=7059&position=P.

- **F/X by Texas Leaguers** (https://pitchfx.texasleaguers.com/): Allows one to set a time frame and find PITCHf/x pitching or batting data for one particular player. This site includes charts on trajectory and movement, tables on pitch characteristics, and outcomes and pitcher/batter match-ups.

- **Prof. Alan Nathan's The Physics of Baseball** (http://baseball.physics.illinois.edu/): Contains research on baseball physics and has a section dedicated to pitch tracking using video technology at http://baseball.physics.illinois.edu/pitchtracker.html.

- **Katron's MLB Gameday BIP Location** (https://katron.org/projects/baseball/hit-location/): Allows one to transpose hit location data of a given ballpark in another ballpark of choice. Keeping in mind all the caveats previously illustrated for batted ball data, it can be used to explore the effect moving to a new team can have on a player's batting.

- **Sportvision**: Sportvision was the company that has devised the PITCHf/x system. They have since been acquired by SMT.

C

Statcast Data Reference

C.1 Introduction

Statcast is the current state-of-the-art tracking system used in all Major League ballparks since the 2015 season. This system is used to track the movements of the baseball and all players on the field at 20,000 frames per second. Using the Statcast system, we can learn about the speed, direction, and distance traveled of players. For example, this system allows for precise evaluation of a defensive player's movement toward a batted ball.

Currently some of the Statcast data is available through the Baseball Savant website, which downloads the data from MLB Advanced Media. The R package **baseballr** has special functions for downloading Statcast pitch-by-pitch data from Baseball Savant. We discuss these in Section C.10. The purpose of this reference is to describe the variables that overlap with variables available in the Retrosheet play-by-play and now defunct PITCHf/x datasets (see Appendix B), and describe the new "off the bat" variables available from Statcast.

C.2 Cross-referencing with Other Data Sources

The `People` table in the **Lahman** database is a useful resource for cross-referencing players across several data sources such as the Baseball-Reference website and the Retrosheet files. Unfortunately, it currently does not contain a column for the MLBAM player identifier; thus the `People` table is not useful for merging Statcast data to information coming from other sources. The best way to cross-reference player identifiers across these systems is by using The Register at Chadwick Baseball Bureau (https://github.com/chadwickbureau/register). There, one finds a link for the download of a zip file containing a register of players, managers, and umpires at any professional level (including, other than the Major Leagues, the Minor and Independent Leagues, Winter Leagues, Japanese and Korean top levels, and the Negro Leagues).

Simpler still is to use the `chadwick_player_lu()` function from the **baseballr** package, as we did in Section 7.5. Since this file takes a minute to download and process, we can store a local copy using the `write_rds()` function.

```
master_id <- baseballr::chadwick_player_lu() |>
  write_rds(
    here::here("data/chadwick_register.rds"), compress = "xz"
  )
```

C.3   Game Situation Variables

Many of the variables concern the game situation at the time of the pitch (see Table C.1). These variables include the date, inning, and number of outs.

TABLE C.1
Game situation variables from Statcast.

Name	Description
game_date	Date of game
batter	Id of the batter
pitcher	Id of the pitcher
stand	Side of the batter
p_throws	Throwing hand of pitcher
home_team	Code for home team
away_team	Code for visiting team
balls	Number of current balls
strikes	Number of current strikes
on_3b	Id of baserunner on third base
on_2b	Id of baserunner on second base
on_1b	Id of baserunner on first base
outs_when_up	Current number of outs
inning	Current inning
inning_topbot	Top or bottom of inning
pos1_person_id	Id of pitcher
pos2_person_id	Id of catcher
pos3_person_id	Id of first baseman
pos4_person_id	Id of second baseman
pos5_person_id	Id of third baseman
pos6_person_id	Id of shortstop
pos7_person_id	Id of left fielder
pos8_person_id	Id of center fielder
pos9_person_id	Id of right fielder
pitch_number	Number of pitch in PA

TABLE C.2
Pitch variables from Statcast.

Name	Description
pitch_type	code for pitch type
pitch_name	pitch type
description	description of outcome of pitch
release_speed	speed of pitch (mph) when released
effective_speed	speed of pitch (mph) when crossing plate
release_pos_x	x-coordinate of release point of pitch
release_pos_y	y-coordinate of release point of pitch
release_pos_z	z-coordinate of release point of pitch
zone	zone location of pitch
pfx_x	horizontal movement of pitch
pfx_z	vertical movement of pitch
sz_top	vertical location of top of strike zone
sz_bot	vertical location of bottom of strike zone
plate_x	horizontal location of pitch
plate_z	vertical location of pitch
vx0	x-coordinate of pitch velocity
vy0	y-coordinate of pitch velocity
vz0	z-coordinate of pitch velocity
ax	x-coordinate of pitch acceleration
ay	y-coordinate of pitch acceleration
az	z-coordinate of pitch acceleration
release_spin_rate	spin rate
spin_axis	spin direction

The identities of all players on the field together with the identities of the baserunners are included. With respect to the specific plate appearance, the dataset includes the pitch number, the number of balls and strikes, and the batting side and throwing hand of the pitcher.

C.4 Pitch Variables

Similar to the PITCHf/x system, this Statcast dataset contains information about each pitch. The variables in Table C.2 include the release point of the pitch, its speed in miles per hour, and movement in the horizontal and vertical directions. The location of the pitch in the zone is recorded and it is classified into a particular region using the zone variable. Using a classification method, the pitch type is recorded. See Table C.3 for the decoding of the abbreviations.

TABLE C.3

The pitch_type and pitch_name variables used by Statcast.

pitch_type	pitch_name
CH	Changeup
CS	Slow Curve
CU	Curveball
EP	Eephus
FA	Other
FC	Cutter
FF	4-Seam Fastball
FO	Forkball
FS	Split-Finger
KC	Knuckle Curve
KN	Knuckleball
PO	Pitch Out
SC	Screwball
SI	Sinker
SL	Slider
ST	Sweeper
SV	Slurve
NA	NA

Here are more detailed descriptions of the pitch variables.

- **release_speed** and **effective_speed**: Speed in miles per hour at the release point and when the ball crosses the front of home plate.

- **sz_top** and **sz_bot**: Vertical coordinates for the top and the bottom of the strike zone of the batter currently at the plate. Both variables are expressed as feet from the ground and they are manually recorded at the beginning of every at-bat.

- **pfx_x** and **pfx_z**: Horizontal and vertical movement of the pitch compared to a theoretical pitch of the same speed with no spin-induced movement. Both variables are measured in inches.

- **plate_x** and **plate_z**: Horizontal and vertical location of the pitch, measured when the pitch crosses the front of home plate. The coordinate system is centered on the middle of home plate and at ground level and viewed from the catcher/umpire point of view, thus a positive value of `plate_x` indicates the pitch crosses the plate to the right of its middle and a negative value to the left. A negative value of `plate_z` indicates a pitch that bounced before reaching home plate. Both `plate_x` and `plate_z` variables are measured in feet.

- **release_pos_x**, **release_pos_y**, **release_pos_z**: Coordinates indicating the calculated position of the ball at the release point. The `release_pos_y` parameter indicates the distance from home plate and is generally set at 50 feet from home plate; researchers have found 55 feet as a distance that better approximates the true release point of the pitch and it is thus advisable to recalculate the coordinates at the 55 foot mark, as illustrated in Section C.5. `release_pos_x`, `release_pos_y`, and `release_pos_z` are the left and right position and the height of the release point in the same coordinate system as `plate_x` and `plate_z`.

- **vx0**, **vy0**, and **vz0**: Components of the pitch velocity in three dimensions, measured at release in feet per second.

- **ax**, **ay**, and **az**: Components of the pitch acceleration in three dimensions, measured at release in ft/s^2.

- **release_spin_rate**: Spin rate of the ball in revolutions per minute.

- **spin_axis**: Direction of the spin of the ball, where $0°$ indicates a perfect top spin and $180°$ indicates a perfect bottom spin.

C.5 Calculating the Pitch Trajectory

As seen in the previous sections, Statcast tracks data on location, velocity, and acceleration of a pitch. Using the kinematics equation for constant acceleration, the position of the ball at a given time t can be determined by the following equations:

$$x = x_0 + xv_0t + \frac{1}{2}axt$$

$$y = y_0 + yv_0t + \frac{1}{2}ayt$$

$$z = z_0 + zv_0t + \frac{1}{2}azt$$

The previous equations are translated to R with use of the following function `pitchloc()`.[1]

```
pitchloc <- function(t, x0, ax, vx0,
                     y0, ay, vy0, z0, az, vz0) {
  x <- x0 + vx0 * t + 0.5 * ax * I(t ^ 2)
  y <- y0 + vy0 * t + 0.5 * ay * I(t ^ 2)
  z <- z0 + vz0 * t + 0.5 * az * I(t ^ 2)
```

---

[1]The code in this section has been slightly adapted from https://code.google.com/p/r-pitchfx/.

```
    if(length(t) == 1) {
      loc <- c(x, y, z)
    } else {
      loc <- cbind(x, y, z)
    }
    return(loc)
}
```

The function `pitch_trajectory()` calculates the trajectory of a pitch from release point to home plate at specified time intervals (the default choice of the argument `interval` is 0.01 seconds).

```
pitch_trajectory <- function(x0, ax, vx0,
                             y0, ay, vy0, z0, az, vz0,
                             interval = 0.01) {
  cross_p <- (-1 * vy0 - sqrt(I(vy0 ^ 2) - 2 * y0 * ay)) / ay
  tracking <- t(
    sapply(
      seq(0, cross_p, interval),
      pitchloc,
      x0 = x0, ax = ax, vx0 = vx0,
      y0 = y0, ay = ay, vy0 = vy0,
      z0 = z0, az = az, vz0 = vz0
    )
  )
  colnames(tracking) <- c("x", "y", "z")
  tracking <- data.frame(tracking)
  return(tracking)
}
```

C.6   Play Event Variables

Although each row of the data set represents a pitch, several variables in Table C.4 record the outcome of the plate appearance. The `type` variable indicates if the ball is a strike, ball, or put in play. The `events`, `des`, and `description` variable provide descriptions of the outcome of the plate appearance.

TABLE C.4
Play event variables.

Name	Description
type	ball or strike or ball in play
events	outcome of plate appearance
des	detailed description of outcome of plate appearance

TABLE C.5
Batted ball variables.

Name	Description
hit_distance_sc	distance away (ft.) that ball lands
hc_x	x location of batted ball when it lands
hc_y	y location of batted ball when it lands
launch_speed	speed of ball as it comes off of the bat
launch_angle	vertical angle at which ball leaves bat
barrel	classification to batted-ball events whose comparable hit types led to a minimum .500 AVG and 1.500 SLG

C.7   Batted Ball Variables

One special aspect of the Statcast dataset is the inclusion of variables about balls that are put into play described in Table C.5. These variables include the exit velocity and launch angle off of the bat, the (x, y)-coordinates of the location of the batted ball, and its estimated distance from home plate. A **barrel** is a way of categorizing a well-hit ball with good combinations of exit velocity and launch angle.

The batted location variables `hc_x` and `hc_y` are related to the spray angle ϕ by the equation

$$\phi = atan\left(\frac{hc_x - 125.42}{198.27 - hc_y}\right).$$

We show this graphically in Figure C.1.

C.8   Derived Variables

Based on the batted ball variables, Statcast has developed several metrics that help in understanding the quality of a specific batted ball, shown in Table C.6. Based on the launch speed and launch angle, one variable `estimated_ba_using_speedangle` gives the estimated probability of a base hit, and a second variable `estimated_woba_using_speedangle` provides the estimate of the weighted on-base percentage for this batted ball.

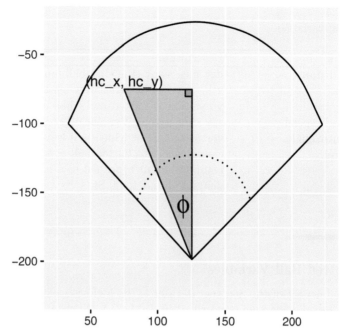

FIGURE C.1
Relationship of Statcast variables hc_x and hc_y with the spray angle ϕ.

TABLE C.6
Statcast derived variables.

Name	Description
estimated_ba_using_speedangle	estimated hit probability
estimated_woba_using_speedangle	estimated woba value

C.9   Defense Variables

Statcast also includes information about the defensive alignments of the teams, shown in Table C.7. The if_fielding_alignment variable indicates if the defensive infield is "standard", "infield shift" (three or more infielders on same side of second base), or "strategic positioning". The of_fielding_alignment can either be "standard", "strategic", or "4th outfielder". Currently, there is some debate about the value of these new defensive alignments and the inclusion of these variables can help determine the effectiveness of these strategies.

TABLE C.7
Statcast defensive alignment variables.

Name	Description
if_fielding_alignment	infield positioning
of_fielding_alignment	outfield positioning

C.10 Acquiring Statcast Data

The `statcast_search()` function from the **baseballr** package will allow you to download Statcast data from Baseball Savant over a specified period of time, or for a particular player. For example, Andrew McCutchen, Freddie Freeman, and José Altuve recorded their 2000th career hits on June 11, June 25, and August 19, 2023, respectively. To retrieve data for McCutchen during the three days before and after his hit, we can use the `statcast_search()` function. There are various ways to find McCutchen's MLB player identifier (see Section C.2), which in this case is 457705.

```
library(baseballr)
mccutchen <- statcast_search(
  start_date = "2023-06-08",
  end_date = "2023-06-14",
  playerid = 457705
)
mccutchen |>
  filter(game_date == "2023-06-11", events == "single") |>
  select(pitch_type, release_speed, release_spin_rate)
```

```
# A tibble: 1 x 3
  pitch_type release_speed release_spin_rate
  <chr>          <dbl>           <dbl>
1 SL             85.8            2502
```

McCutchen's 2000th hit came off of an 86 mph slider spinning at 2500 revolutions per minute.

Please see Section 12.2 for information about how to store one or more years of Statcast data.

References

Adler, Joseph. 2006. *Baseball Hacks: Tips & Tools for Analyzing and Winning with Statistics.* Sebastopol, CA: O'Reilly Media.

Albert, Jim. 2002. "Smoothing Career Trajectories of Baseball Hitters." Bowling Green State University.

———. 2008. "Streaky Hitting in Baseball." *Journal of Quantitative Analysis in Sports* 4 (1). https://doi.org/10.2202/1559-0410.1085.

———. 2009. "Is Roger Clemens' WHIP Trajectory Unusual?" *Chance* 22 (2): 8–20. https://doi.org/10.1080/09332480.2009.10722954.

———. 2017. *Teaching Statistics Using Baseball.* Washington, DC: Mathematical Association of America.

———. 2018. *LearnBayes: Functions for Learning Bayesian Inference.* https://CRAN.R-project.org/package=LearnBayes.

Albert, Jim, Jay Bartroff, Roger Blandford, Dan Brooks, Josh Derenski, Larry Goldstein, Hosoi Anette, Gary Lorden, Alan Nathan, and Lloyd Smith. 2018. "Report of the Committee Studying Home Run Rates in MLB." http://baseball.physics.illinois.edu/HRReport2018.pdf.

Albert, Jim, and Jay Bennett. 2003. *Curve Ball: Baseball, Statistics, and the Role of Chance in the Game.* New York: Copernicus Books.

Albert, Jim, Anette Hosoi, Alan Nathan, and Lloyd Smith. 2019. "Preliminary Report of the Committee Studying Home Run Rates in MLB." http://baseball.physics.illinois.edu/HRReport2019.pdf.

Albert, Jim, and Alan Nathan. 2022. "Home Runs and Drag: An Early Look at the 2022 Season." Fangraphs. https://blogs.fangraphs.com/home-runs-and-drag-an-early-look-at-the-2022-season/.

Albert, Jim, and Maria Rizzo. 2012. *R by Example.* New York: Springer Science & Business Media.

Allaire, J. J., Yihui Xie, Christophe Dervieux, Jonathan McPherson, Javier Luraschi, Kevin Ushey, Aron Atkins, et al. 2023. *rmarkdown: Dynamic Documents for r.* https://github.com/rstudio/rmarkdown.

Allen, Dave. 2009a. "Deconstructing the Non-Fastball Run Maps." Baseball Analysts. http://baseballanalysts.com/archives/2009/03/deconstructing_1.php.

———. 2009b. "Platoon Splits for Three Types of Fastballs." Baseball Analysts. http://baseballanalysts.com/archives/2009/05/platoon_splits.php.

Appelman, David. 2008. "Get to Know: RE24." Fangraphs. http://www.fangraphs.com/blogs/get-to-know-re24/.

Bates, Douglas, Martin Mächler, Ben Bolker, and Steve Walker. 2015. "Fitting Linear Mixed-Effects Models Using lme4." *Journal of Statistical Software* 67 (1): 1–48. https://doi.org/10.18637/jss.v067.i01.

Baumer, Benjamin S., and Jim Albert. 2024. *Abdwr3edata: Companion to "Analyzing Baseball Data with R," 3rd Edition.* https://github.com/beanu mber/abdwr3edata.

Baumer, Benjamin S., Shane T. Jensen, and Gregory J. Matthews. 2015. "open-WAR: An Open Source System for Evaluating Overall Player Performance in Major League Baseball." *Journal of Quantitative Analysis in Sports* 11 (2): 69–84. https://doi.org/10.1515/jqas-2014-0098.

Baumer, Benjamin S., Daniel T. Kaplan, and Nicholas J. Horton. 2021a. *Modern Data Science with R.* 2nd ed. Boca Raton: Chapman; Hall/CRC Press. https://www.routledge.com/Modern-Data-Science-with-R/Baumer-Kaplan-Horton/p/book/9780367191498.

———. 2021b. *Modern Data Science with R.* 2nd ed. Boca Raton, FL: Chapman; Hall/CRC Press. https://www.routledge.com/Modern-Data-Science-with-R/Baumer-Kaplan-Horton/p/book/9780367191498.

Baumer, Benjamin S., Gregory J. Matthews, and Quang Nguyen. 2023. "Big Ideas in Sports Analytics and Statistical Tools for Their Investigation." *Wiley Interdisciplinary Reviews: Computational Statistics*, e1612. https://doi.org/10.1002/wics.1612.

Berry, Scott M. 1991. "The Summer of '41: A Probabilistic Analysis of DiMaggio's "Streak" and Williams's Average of .406." *Chance* 4 (4): 8–11. https://doi.org/10.1080/09332480.1991.10542337.

Berry, Scott M., C. Shane Reese, and Patrick D. Larkey. 1999. "Bridging Different Eras in Sports." *Journal of the American Statistical Association* 94 (447): 661–76. https://doi.org/10.1080/01621459.1999.10474163.

Bouzarth, Elizabeth, Benjamin Grannan, John Harris, Andrew Hartley, Kevin Hutson, and Ella Morton. 2021. "Swing Shift: A Mathematical Approach to Defensive Positioning in Baseball." *Journal of Quantitative Analysis in Sports* 17 (1): 47–55. https://doi.org/10.1515/jqas-2020-0027.

Bradley, Ralph Allan, and Milton E. Terry. 1952. "Rank Analysis of Incomplete Block Designs: I. The Method of Paired Comparisons." *Biometrika* 39 (3/4): 324–45. https://doi.org/10.2307/2334029.

Brill, Ryan S., Sameer K. Deshpande, and Abraham J. Wyner. 2023. "A Bayesian Analysis of the Time Through the Order Penalty in Baseball." *Journal of Quantitative Analysis in Sports*, no. 0. https://doi.org/10.1515/jqas-2022-0116.

Brooks, Dan, Harry Pavilidis, and Jonathan Judge. 2015. "Moving Beyond WOWY: A Mixed Approach to Measuring Catcher Framing." Baseball Prospectus. https://www.baseballprospectus.com/news/article/25514/moving-beyond-wowy-a-mixed-approach-to-measuring-catcher-framing/.

Brooks, Dan, and Harry Pavlidis. 2014. "Framing and Blocking Pitches: A Regressed, Probabilistic Model: A New Method for Measuring Catcher Defense." Baseball Prospectus. https://www.baseballprospectus.com/new

s/article/22934/framing-and-blocking-pitches-a-regressed-probabilistic-model-a-new-method-for-measuring-catcher-defense/.

Bukiet, Bruce, Elliotte Rusty Harold, and José Luis Palacios. 1997. "A Markov Chain Approach to Baseball." *Operations Research* 45 (1): 14–23. https://doi.org/10.1287/opre.45.1.14.

Campitelli, Elio. 2021. *metR: Tools for Easier Analysis of Meteorological Fields.* https://doi.org/10.5281/zenodo.2593516.

Caola, Ralph. 2003. "Using Calculus to Relate Runs to Wins: Part i." *By the Numbers* 13: 9–16.

Carl, Sebastian, and Camden Kay. 2023. *mlbplotR: Create "ggplot2" and "gt" Visuals with Major League Baseball Logos.* https://CRAN.R-project.org/package=mlbplotR.

Casals, Martí, José Fernández, Victor Martínez, Michael Lopez, Klaus Langohr, and Jordi Cortés. 2023. "A Systematic Review of Sport-Related Packages Within the r CRAN Repository." *International Journal of Sports Science & Coaching* 18 (2): 621–29. https://doi.org/10.1177/17479541221136238.

Chang, Winston, Joe Cheng, J. J. Allaire, Carson Sievert, Barret Schloerke, Yihui Xie, Jeff Allen, Jonathan McPherson, Alan Dipert, and Barbara Borges. 2023. *shiny: Web Application Framework for r.* https://CRAN.R-project.org/package=shiny.

Cleveland, William S. 1979. "Robust Locally Weighted Regression and Smoothing Scatterplots." *Journal of the American Statistical Association* 74 (368): 829–36. https://doi.org/10.1080/01621459.1979.10481038.

———. 1985. *The Elements of Graphing Data.* Vol. 2. Monterey, CA: Wadsworth Advanced Books; Software.

Csárdi, Gábor, Jim Hester, Hadley Wickham, Winston Chang, Martin Morgan, and Dan Tenenbaum. 2023. *remotes: R Package Installation from Remote Repositories, Including "GitHub".* https://CRAN.R-project.org/package=remotes.

Dahl, David B., David Scott, Charles Roosen, Arni Magnusson, and Jonathan Swinton. 2019. *xtable: Export Tables to LaTeX or HTML.* https://CRAN.R-project.org/package=xtable.

Davenport, Clay, and Keith Woolner. 1999. "Revisiting the Pythagorean Theorem." Baseball Prospectus. www.baseballprospectus.com/ article.php?articleid=342.

Deshpande, Sameer K., and Abraham Wyner. 2017. "A Hierarchical Bayesian Model of Pitch Framing." *Journal of Quantitative Analysis in Sports* 13 (3): 95–112. https://doi.org/10.1515/jqas-2017-0027.

Donoho, David. 2017. "50 Years of Data Science." *Journal of Computational and Graphical Statistics* 26 (4): 745–66. https://doi.org/10.1080/10618600.2017.1384734.

Douglas, Colin and Richard Scriven. 2024. retrosheet: Import Professional Baseball Data from 'Retrosheet'. R package version 1.1.6. https://CRAN.Rproject.org/package=retrosheet.

Fair, Ray C. 2008. "Estimated Age Effects in Baseball." *Journal of Quantitative Analysis in Sports* 4 (1). https://doi.org/10.2202/1559-0410.1074.

Fast, Mike. 2010. "What the Heck Is PITCHf/x." *The Hardball Times Annual* 2010: 153–58.

———. 2011. "Spinning Yarn: Removing the Mask Encore Presentation." Baseball Prospectus. https://www.baseballprospectus.com/news/article/15093/spinning-yarn-removing-the-mask-encore-presentation/.

Francisco Rodriguez-Sanchez, and Connor P. Jackson. 2023. *grateful: Facilitate Citation of r Packages*. https://pakillo.github.io/grateful/.

Friendly, Michael, Chris Dalzell, Martin Monkman, and Dennis Murphy. 2023. *Lahman: Sean "Lahman" Baseball Database*. https://CRAN.R-project.org/package=Lahman.

Gerber, Eric A. E., and Bruce A. Craig. 2021. "A Mixed Effects Multinomial Logistic-Normal Model for Forecasting Baseball Performance." *Journal of Quantitative Analysis in Sports* 17 (3): 221–39. https://doi.org/10.1515/jqas-2020-0007.

Gould, Stephen Jay. 1989. "The Streak of Streaks." *Chance* 2 (2): 10–16. https://doi.org/10.1080/09332480.1989.10554932.

Grolemund, Garrett, and Hadley Wickham. 2011. "Dates and Times Made Easy with lubridate." *Journal of Statistical Software* 40 (3): 1–25. https://www.jstatsoft.org/v40/i03/.

Harrell Jr, Frank E. 2023. *Hmisc: Harrell Miscellaneous*. https://CRAN.R-project.org/package=Hmisc.

Healey, Glenn. 2019. "A Bayesian Method for Computing Intrinsic Pitch Values Using Kernel Density and Nonparametric Regression Estimates." *Journal of Quantitative Analysis in Sports* 15 (1): 59–74. https://doi.org/10.1515/jqas-2017-0058.

Heipp, B. 2003. "W% Estimators." Buckeyes and Sabermetrics. http://gosu02.tripod.com/id69.html.

Hester, Jim, and Davis Vaughan. 2023. *bench: High Precision Timing of r Expressions*. https://CRAN.R-project.org/package=bench.

Hester, Jim, Hadley Wickham, and Gábor Csárdi. 2023. *fs: Cross-Platform File System Operations Based on "libuv"*. https://CRAN.R-project.org/package=fs.

Hirotsu, Nobuyoshi, and J. Eric Bickel. 2019. "Using a Markov Decision Process to Model the Value of the Sacrifice Bunt." *Journal of Quantitative Analysis in Sports* 15 (4): 327–44. https://doi.org/10.1515/jqas-2017-0092.

Ismay, Chester, and Albert Y. Kim. 2019. *Modern Dive: Statistical Inference via Data Science*. Boca Raton, FL: CRC Press. https://moderndive.com/.

James, Bill. 1980. *Baseball Abstract*. Lawrence, KS: self-published.

———. 1982. *Baseball Abstract*. New York: Ballantine Books.

———. 1994. *The Politics of Glory: How Baseball's Hall of Fame Really Works*. London: Macmillan.

Judge, Jonathan. 2018. "Bayesian Bagging to Generate Uncertainty Intervals: A Catcher Framing Story." Baseball Prospectus. https://www.baseballpros

pectus.com/news/article/38289/bayesian-bagging-generate-uncertainty-intervals-catcher-framing-story/.

Kabacoff, Robert I. 2010. *R in Action*. New York: Manning Publications.

Kemeny, John G., and James Laurie Snell. 1960. *Finite Markov Chains*. Vol. 210. New York: Springer-Verlag.

Keri, Jonah, and Baseball Prospectus. 2007. *Baseball Between the Numbers: Why Everything You Know about the Game Is Wrong*. New York: Basic Books.

Lahman, Sean. 2018. "Lahman's Baseball Database, 1871–2017." seanlahman.com. http://seanlahman.com/.

Lindbergh, Ben. 2013. "The Art of Pitch Framing." Grantland. http://grantland.com/features/studying-art-pitch-framing-catchers-such-francisco-cervelli-chris-stewart-jose-molina-others/.

Lindsey, George R. 1963. "An Investigation of Strategies in Baseball." *Operations Research* 11 (4): 477–501. https://doi.org/10.1287/opre.11.4.477.

Lopez, Michael J., Gregory J. Matthews, and Benjamin S. Baumer. 2018. "How Often Does the Best Team Win? A Unified Approach to Understanding Randomness in North American Sport." *The Annals of Applied Statistics* 12 (4): 2483–2516. https://doi.org/10.1214/18-AOAS1165.

Marchi, Max. 2010. "Platoon Splits 2.0." Hardball Times. http://www.hardballtimes.com/main/article/platoon-splits-2.0.

McCotter, Trent. 2010. "Hitting Streaks Don't Obey Your Rules: Evidence That Hitting Streaks Are Not Just Byproducts of Random Variation." *Chance* 23 (4): 52–57. https://doi.org/10.1080/09332480.2010.10739837.

Meschiari, Stefano. 2022. *Latex2exp: Use LaTeX Expressions in Plots*. https://CRAN.R-project.org/package=latex2exp.

Mühleisen, Hannes, and Mark Raasveldt. 2023. *duckdb: DBI Package for the DuckDB Database Management System*. https://CRAN.R-project.org/package=duckdb.

Müller, Kirill. 2020. *here: A Simpler Way to Find Your Files*. https://CRAN.R-project.org/package=here.

Müller, Kirill, Jeroen Ooms, David James, Saikat DebRoy, Hadley Wickham, and Jeffrey Horner. 2023. *RMariaDB: Database Interface and MariaDB Driver*. https://CRAN.R-project.org/package=RMariaDB.

Müller, Kirill, Hadley Wickham, David A. James, and Seth Falcon. 2024. *RSQLite: SQLite Interface for r*. https://CRAN.R-project.org/package=RSQLite.

Murrell, Paul. 2006. *R Graphics*. Boca Raton, FL: Chapman & Hall, CRC Press.

Nathan, Alan M. 2011. "Baseball ProGUESTus: Home Runs and Humidors: Is There a Connection?" Baseball Prospectus. https://www.baseballprospectus.com/news/article/13057/baseball-proguestus-home-runs-and-humidors-is-there-a-connection/.

Official Playing Rules Committee. 2018. *2018 Official Rules of Major League Baseball*. Chicago, IL: Triumph Books. http://www.triumphbooks.com/201 8-official-rules-of-major-league-baseball-products-9781629375434.php.

Palmer, Pete. 1983. "Balls and Strikes." Baseball Analyst. http://sabr.org/res earch/baseball-analyst-archives.

Pankin, Mark D. 1987. "Baseball as a Markov Chain." In *The Great American Baseball Stat Book*, 1st ed., 520–24. New York: Ballantine Books.

Pedersen, Thomas Lin. 2024. *patchwork: The Composer of Plots*. https://CR AN.R-project.org/package=patchwork.

Petti, Bill, and Saiem Gilani. 2024. *baseballr: Acquiring and Analyzing Baseball Data*. https://CRAN.R-project.org/package=baseballr.

R Core Team. 2023. *R: A Language and Environment for Statistical Computing*. Vienna, Austria: R Foundation for Statistical Computing. https://www.R-project.org/.

Richardson, Neal, Ian Cook, Nic Crane, Dewey Dunnington, Romain François, Jonathan Keane, Dragoș Moldovan-Grünfeld, Jeroen Ooms, Jacob Wujciak-Jens, and Apache Arrow. 2023. *arrow: Integration to "Apache" "Arrow"*. https://CRAN.R-project.org/package=arrow.

Robinson, David, Alex Hayes, and Simon Couch. 2023. *broom: Convert Statistical Objects into Tidy Tibbles*. https://CRAN.R-project.org/package=br oom.

RStudio Team. 2018. *RStudio: Integrated Development Environment for R*. Boston, MA: RStudio, Inc. http://www.rstudio.com/.

Schwarz, Alan. 2004. *The Numbers Game: Baseball's Lifelong Fascination with Statistics*. New York: St. Martin's Press.

Seidel, Michael. 2002. *Streak: Joe DiMaggio and the Summer of '41*. Lincoln, NE: University of Nebraska Press.

Slowikowski, Kamil. 2024. *ggrepel: Automatically Position Non-Overlapping Text Labels with "ggplot2"*. https://CRAN.R-project.org/package=ggrepel.

Star, J. 2011. "The Road to October: Sept. 29, https://www.nytimes.com/20 11/09/30/sports/baseball/5-hour-joy-ride-like-no-other.html.

Tango, Tom M., Mitchel G. Lichtman, and Andrew E. Dolphin. 2007. *The Book: Playing the Percentages in Baseball*. Dulles, VA: Potomac Books, Inc.

Turkenkopf, Dan. 2008. "Framing the Debate." Beyond the Box Score. https://www.beyondtheboxscore.com/2008/4/5/389840/framing-the-debate.

Venables, W. N., D. M. Smith, and the R Development Core Team. 2011. "An Introduction to R: Notes on R, a Programming Environment for Data Analysis and Graphics, v. 2.13.0."

Walsh, John. 2008. "Searching for the Game's Best Pitch." The Hardball Times. http://www.hardballtimes.com/main/article/searching-for-the-games-best-pitch/.

———. 2010. "The Compassionate Umpire." The Hardball Times. http://www.hardballtimes.com/main/article/the-compassionate-umpire/.

Waring, Elin, Michael Quinn, Amelia McNamara, Eduardo Arino de la Rubia, Hao Zhu, and Shannon Ellis. 2022. *skimr: Compact and Flexible Summaries of Data.* https://CRAN.R-project.org/package=skimr.

Wickham, Hadley. 2014. "Tidy Data." *Journal of Statistical Software* 59 (10): 1–23. https://doi.org/10.18637/jss.v059.i10.

———. 2016b. *ggplot2: Elegant Graphics for Data Analysis.* New York: Springer.

———. 2016a. *Ggplot2: Elegant Graphics for Data Analysis.* New York: Springer-Verlag. https://ggplot2.tidyverse.org.

———. 2022a. *lobstr: Visualize r Data Structures with Trees.* https://CRAN.R-project.org/package=lobstr.

———. 2022b. *rvest: Easily Harvest (Scrape) Web Pages.* https://CRAN.R-project.org/package=rvest.

———. 2023a. *downlit: Syntax Highlighting and Automatic Linking.* https://CRAN.R-project.org/package=downlit.

———. 2023b. *modelr: Modelling Functions That Work with the Pipe.* https://CRAN.R-project.org/package=modelr.

———. 2023c. *stringr: Simple, Consistent Wrappers for Common String Operations.* https://CRAN.R-project.org/package=stringr.

Wickham, Hadley, Mara Averick, Jennifer Bryan, Winston Chang, Lucy D'Agostino McGowan, Romain François, Garrett Grolemund, et al. 2019. "Welcome to the tidyverse." *Journal of Open Source Software* 4 (43): 1686. https://doi.org/10.21105/joss.01686.

Wickham, Hadley, Mine Çetinkaya-Rundel, and Garrett Grolemund. 2023. *R for Data Science.* 2nd ed. Sebastapol, CA: O'Reilly Media, Inc. https://r4ds.hadley.nz/.

Wickham, Hadley, Romain François, Lionel Henry, Kirill Müller, and Davis Vaughan. 2023. *dplyr: A Grammar of Data Manipulation.* https://CRAN.R-project.org/package=dplyr.

Wickham, Hadley, Maximilian Girlich, and Edgar Ruiz. 2023. *dbplyr: A "dplyr" Back End for Databases.* https://CRAN.R-project.org/package=dbplyr.

Wickham, Hadley, Jim Hester, and Jennifer Bryan. 2024. *readr: Read Rectangular Text Data.* https://CRAN.R-project.org/package=readr.

Wickham, Hadley, Jim Hester, and Jeroen Ooms. 2023. *Xml2: Parse XML.* https://CRAN.R-project.org/package=xml2.

Wilkinson, Leland. 2006. *The Grammar of Graphics.* New York: Springer Science & Business Media.

Wood, S. N. 2017. *Generalized Additive Models: An Introduction with r.* 2nd ed. Chapman; Hall/CRC.

Wood, S. N. 2003. "Thin-Plate Regression Splines." *Journal of the Royal Statistical Society (B)* 65 (1): 95–114.

———. 2004. "Stable and Efficient Multiple Smoothing Parameter Estimation for Generalized Additive Models." *Journal of the American Statistical Association* 99 (467): 673–86.

————. 2011. "Fast Stable Restricted Maximum Likelihood and Marginal Likelihood Estimation of Semiparametric Generalized Linear Models." *Journal of the Royal Statistical Society (B)* 73 (1): 3–36.

Wood, S. N., N. Pya, and B. Säfken. 2016. "Smoothing Parameter and Model Selection for General Smooth Models (with Discussion)." *Journal of the American Statistical Association* 111: 1548–75.

Xie, Yihui. 2014. "knitr: A Comprehensive Tool for Reproducible Research in R." In *Implementing Reproducible Computational Research*, edited by Victoria Stodden, Friedrich Leisch, and Roger D. Peng. Chapman; Hall/CRC.

————. 2015. *Dynamic Documents with R and Knitr*. 2nd ed. Boca Raton, Florida: Chapman; Hall/CRC. https://yihui.org/knitr/.

————. 2023. *knitr: A General-Purpose Package for Dynamic Report Generation in r*. https://yihui.org/knitr/.

Xie, Yihui, J. J. Allaire, and Garrett Grolemund. 2018. *R Markdown: The Definitive Guide*. Boca Raton, Florida: Chapman; Hall/CRC. https://bookdown.org/yihui/rmarkdown.

Xie, Yihui, Christophe Dervieux, and Emily Riederer. 2020. *R Markdown Cookbook*. Boca Raton, Florida: Chapman; Hall/CRC. https://bookdown.org/yihui/rmarkdown-cookbook.

Zeileis, Achim, and Gabor Grothendieck. 2005. "zoo: S3 Infrastructure for Regular and Irregular Time Series." *Journal of Statistical Software* 14 (6): 1–27. https://doi.org/10.18637/jss.v014.i06.

Zhu, Hao. 2024. *kableExtra: Construct Complex Table with "kable" and Pipe Syntax*. https://CRAN.R-project.org/package=kableExtra.

Indices

The Subject Index catalogs general items of interest, including sabermetric concepts, people, teams, and other notions generally found in an index.

The R Index contains all R functions mentioned in the text. Packages are listed as subitems under the `library()` function.

Subject index

R index

For Product Safety Concerns and Information please contact our
EU representative GPSR@taylorandfrancis.com Taylor & Francis
Verlag GmbH, Kaufingerstraße 24, 80331 München, Germany